Springer Aerospace Technology

The *Springer Aerospace Technology* series is devoted to the technology of aircraft and spacecraft including design, construction, control and the science. The books present the fundamentals and applications in all fields related to aerospace engineering. The topics include aircraft, missiles, space vehicles, aircraft engines, propulsion units and related subjects.

More information about this series at http://www.springer.com/series/8613

Sauta O.I. · Shatrakov A.Y. ·
Shatrakov Y.G. · Zavalishin O.I.

Principles of Radio Navigation for Ground and Ship-Based Aircrafts

Sauta O.I.
Saint-Petersburg, Russia

Shatrakov Y.G.
Saint-Petersburg, Russia

Shatrakov A.Y.
Moscow, Russia

Zavalishin O.I.
Moscow, Russia

ISSN 1869-1730 ISSN 1869-1749 (electronic)
Springer Aerospace Technology
ISBN 978-981-13-8295-6 ISBN 978-981-13-8293-2 (eBook)
https://doi.org/10.1007/978-981-13-8293-2

This Springer imprint is published by the registered company Springer Nature Singapore Pte Ltd.
The registered company address is: 152 Beach Road, #21-01/04 Gateway East, Singapore 189721, Singapore

Introduction

Aviation capabilities largely depend on accuracy and fidelity of navigation issues solving, state of the art of navigation instruments, and especially radio navigation equipment and systems.

Radio navigation systems (RNS) are designed to determine the coordinates, speed and direction of movement, deviation from a given route, and other parameters of the aircraft, regardless of time and weather conditions. RNS are used on all aircraft types, regardless of their type and purpose.

All considered in the learning guide RNS include at least two functionally interrelated complexes of radio technic equipment:

- Broadcasting ground stations (GS), installed on fixed or mobile objects with known coordinates;
- Interrogators installed on aircraft and allowing to determine the necessary navigation parameters by the received signals from the broadcasting stations.

Some RNS are also characterized by the presence of a third complex—GS control equipment.

RNS, unlike other radio facilities, are measuring systems. Useful information in them is formed not in the transmitting path, but in the path of radio wave propagation, due to the dependence of the parameters of the electromagnetic field of the received signals from the place of the moving object. In fact, the work of RNS is based on the ability of radio waves to spread in a homogeneous medium to the shortest distances with a terminal speed of approximately $3 \cdot 10^5$ km/s.

In this work, widely used in practice RNS are considered. The authors describe in detail the principle of operation, composition, and structure of the equipment, as well as the features of the RNS operation, which is necessary for students of radio engineering specialties. The manual will be useful also for graduate students, who prepare a thesis on the “radar ranging and radio navigation” specialty.

Contents

1 Basic Principles of Radio Navigation 1
1.1 Summary of Radio Navigation Development 1
1.2 Concept of Navigation and Radio Navigation 2
1.3 Coordinate Systems 5
References 9

2 The Concept of Radio Navigation Systems 11
2.1 Features of Radio Waves Propagation 11
2.2 Classification of Radio Navigation Systems 12
2.3 Methods for Determining the Location of Moving Objects 13
2.4 System Working Area 24
2.5 Time Keeping in Radio Navigation Systems 26
References 30

3 Short-Range Navigation Systems 31
3.1 Short-Range Radio Technical Navigation System 31
3.2 TACAN Radio Navigation System 33
3.3 Radio Navigation System VOR/DME 34
3.4 GRAS-Automated System 38
References 42

4 Systems of Long-Range Radio Navigation with Ground-Based Stations 43
4.1 Phase Code Systems of the Long-Wave Range 43
4.2 Super-Long-Wave Range Systems 53
References 64

5 Instrumental Landing Systems 65
5.1 Systems of One Meter Range 65
5.2 Systems of the Decimeter Wave Range 67

5.3 Microwave Systems 69
5.4 Prospects for the Development of Instrumental Landing Systems 71
References 71

6 Satellite Radio Navigation Systems 73
6.1 Overview of the Development of the Systems 73
6.2 Principles of System Operations 74
6.3 Main Characteristics of Systems 83
References 87

7 Differential and Relative Operation Modes of Systems 89
7.1 Principle of Systems Operation in Differential Mode 89
7.2 Principle of Operation in Relative Mode 93
References 97

8 Augmentation Systems to Ground-Based GNSS 99
8.1 Principles of Creation of Satellite Augmentation Systems 99
8.2 General Information on GBAS 100
8.3 Purpose, Basic Requirements and Functions of GBAS 104
8.4 Technical Requirements and Characteristics of LAAS 105
8.5 The Concept of System Data Integrity 108
8.6 Characteristics of Coordinates Determination Accuracy 109
8.7 Monitoring Errors in Ephemeris and GNSS Failures 111
8.8 Continuity of Service 112
8.9 Selecting the Channel and Identifying the LAAS 113
8.10 Final Approach Path 114
8.11 Accounting of Airport Location Conditions 117
8.12 Lateral and Vertical Thresholds of Alarm Actuation 118
8.13 Monitoring and Actions to Keep the System in Good Operating Condition 119
8.14 Accuracy of Setting Reference Parameters 120
8.15 Introduction of GBAS Systems in Russia 120
References 124

9 Automatic-Dependent Surveillance 125
References 130

Abbreviations

ABAS	Aircraft-based augmentation system
AC	Aircraft
ADB	Aircraft database
ADF	Automatic direction finder
ADS-B	Automatic-dependent surveillance—broadcast
AFS	Antenna-feed system
AMS	Airborne multifunctional system
CAS	Collision avoidance system
DC	Differential corrections
DL	Data link
FAS	Final approach segment
FM	Frequency modulated
FS	Flight safety
GBAS	GNSS ground-based augmentation system
GISR	Ground-integrated signal-to-noise ratio
GLONASS	GLObal NAvigation Satellite System (Russia)
GLS	GBAS landing system
GNSS	Global Navigation Satellite System (currently including GLONASS and GPS)
GNSS/LAAS	Onboard satellite-based landing equipment
GPS	Global Positioning System (USA)
GS	Ground stations
IAC	Interstate Aviation Committee
ICAO	International Civil Aviation Organization
ILS	Instrument landing system
LAAS	Local monitoring and correcting station
LAL	Lateral alert limit
LPL	Lateral protection level
NPA	"non-precision" approach
PR	Pseud-orange

RNS Radio navigation systems
RNT Radio navigation points
RR GNSS reference receiver
RWY Runway
SBAS GNSS satellite-based augmentation system
SNS Satellite navigation system
SLS Satellite-based landing system
SV Space vehicle
UTC Coordinated Universal Time
VDB VHF band digital signal for data transmission
VOR VHF omnidirectional radio range

Chapter 1
Basic Principles of Radio Navigation

1.1 Summary of Radio Navigation Development

The development of radio navigation systems began with the invention of radio direction finding apparatus in the early twentieth century, when moving objects, including ships and then aircraft, were able to receive the direction of motion to the nondirectional radio beacons. These stations (radio beacons) were usually installed in the area of final points of movement and were, in fact, analogs of optical beacons, but with a larger range of operation and less dependence on weather conditions. Then, closer to the middle of the twentieth century, there were automatic radio compasses. In parallel with the development of radio technic, engineering thought continued to create new radio navigation systems (RNS) and develop new principles for determining the parameters of the state of moving objects in space.

In the middle of the twentieth century, RNS for short-range navigation are appearing, the working area of which extends to a distance of several hundred kilometers from the ground station, and that allows the aircraft to lay the route through the specified points, as well as to obtain information about the distance (length) to these points of the route and their azimuth [1].

Simultaneously with the development of the RNS for short-range navigation, long-range navigation RNS with support stations located in the points of the globe spaced over long distances are developed and put into operation. Taking into account the frequency range, these systems are divided into systems operating in areas of 3–5 thousand km (frequency 100 kHz) and global action systems that allow to determine the coordinates of mobile objects at any point in the near-Earth space and in the waters of the oceans at a depth up to 15 m. Such RNS operate at frequencies of about 10 kHz [2].

To ensure the approach of the aircraft for landing, which is especially important in difficult weather conditions, in the 50s of the twentieth century, the RNS of instrumental landing are created. The working area of such RNS is limited by the range of several tens of kilometers from the runway of the airport and a narrow sector in the corner of the place and the azimuth relative to the runway axis [3].

Sauta O.I. et al., *Principles of Radio Navigation for Ground and Ship-Based Aircrafts*,
Springer Aerospace Technology, https://doi.org/10.1007/978-981-13-8293-2_1

With the near-Earth space development in the mid-70s of the twentieth century, it became possible to create RNS of global operation, the ground stations of which were placed on spacecrafts. In these RNS, navigation information on the aircraft board is obtained by processing signals from several navigation satellites moving in fixed orbits around the Earth [4].

At the end of the twentieth century, due to the rapid development of airborne and ground avionics, it became possible to transmit the current parameters of the aircraft state (coordinates, speed, etc.) in the air traffic control system (ATC system) and to the other air traffic participants. This has opened up new opportunities for increasing the efficiency of aircraft using and improving flight safety, especially for ship-based aircraft or those operating at poorly equipped airfields and landing sites.

1.2 Concept of Navigation and Radio Navigation

Navigation is the science of the means and methods of determining the coordinates of the movement of various objects [5]. Historically, navigation has emerged and developed as a science in the driving of ships. Nowadays, it is being developed in relation to the movement of any objects. The main purpose of navigation is to ensure movement along a given trajectory in order to reach a given point in space and at a given time. At the same time, some specific tasks are also solved by navigation: determining the location of the object in the defined coordinate system, course, speed, acceleration of movement, the time required to enter a given area, issuing recommendations on the necessary maneuvers, etc. The science that solves the problem of navigation using radio engineering is understood by radio navigation [6]. Nowadays, radio navigation tools play a major role in providing the aircraft with information about the location; these tools, in particular, provide correction of onboard inertial systems, widely used in aviation.

The basic concepts of navigation

In navigation, the track line is called the projection of the trajectory of the object on the surface of the globe. A route is a track line plotted on a map. The true rate of K_i-angle is measured in the horizontal plane between the meridian direction at the point of the moving object and the direction of its longitudinal axis [7].

Magnetic course K_M is the heading angle measured relative to the magnetic meridian direction at the location of the object, measured with a magnetic compass. For determining the true K course, it is necessary to add an amendment on the magnetic declination of ΔM, the value of which depends on the geographical location of the moving object to the value of the magnetic K_M heading

$$K = K_M + \Delta M. \tag{1.1}$$

Bearing P is the angle measured in the horizontal plane between the directions to the North and the bearing object.

The heading angle q is the angle measured in the horizontal plane, between the direction of the longitudinal axis of the moving object and the direction of the bearing object.

The bearing is defined as the sum of the heading angles and the heading angle

$$P = K + q. \tag{1.2}$$

Airspeed V_r is the speed of the aircraft relative to the air masses. The total speed vector of the aircraft relative to the Earth is determined by the sum of the airspeed vectors V_r and wind speed U_r

$$W_r = V_r + U_r \tag{1.3}$$

The projections of these vectors on the horizontal plane form a triangle of velocities (Fig. 1.1).

The horizontal component of the full speed vector of the aircraft is called the ground speed. Ground speed determines the distance traveled above ground by aircraft and is, therefore, very important in navigation.

Track angle α_T—the angle in the horizontal plane between the North direction and the direction of the ground speed vector.

Crab angle α_c—the horizontal angle between the projection of the longitudinal axis of the aircraft and the ground speed vector.

The aerodynamic slip angle α_{sk}—the angle between the horizontal projection of the longitudinal axis of the aircraft and the vector V_r.

The count of all abovementioned angles is clockwise.

When the aircraft is moving, it is usually moved "sideways". This happens due to the wind demolition of the aircraft and aerodynamic slip in the lateral direction when rolling.

The direction of movement is determined by the track angle:

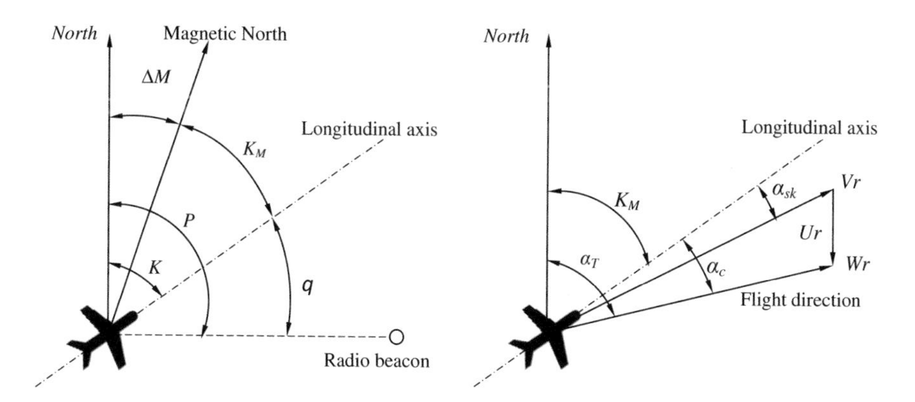

Fig. 1.1 Scheme of angles determining the position of the moving object in space

$$\alpha_T = K + \alpha_c. \quad (1.4)$$

Wind angle is the difference between wind direction and travel angle.

Figure 1.1 shows the designation of mentioned angles and speed vectors.

Most of the flight time, aircraft movement is carried out along trajectories that can be represented through certain geometric relations. The greatest use is in aviation to find the trajectories of geodetic or rhumb lines.

Great circle is an arc of a great circle and a great circle is a circle that is obtained by the crossing of the surface of the Earth by a plane passing through the center of the Earth.

The great circle is the shortest distance line between two points on the Earth's surface (or above it) and is used for laying a path over long distances. Great circle crosses the meridians at different angles (except in the case of moving along the equator). Therefore, the track angle during the movement of the aircraft is continuously changing.

Rhumb line is a line of the path, where the movement vector crosses all current meridians at the same angle. Rhumb line is frequently used for routing the paths in a relatively short distance.

Figure 1.2 schematically shows the arrangement of the geodetic and rhumb lines against the geodetic coordinate system.

When performing aircraft flights many local and global coordinate systems are used to solve specific navigation tasks at different stages of the flight.

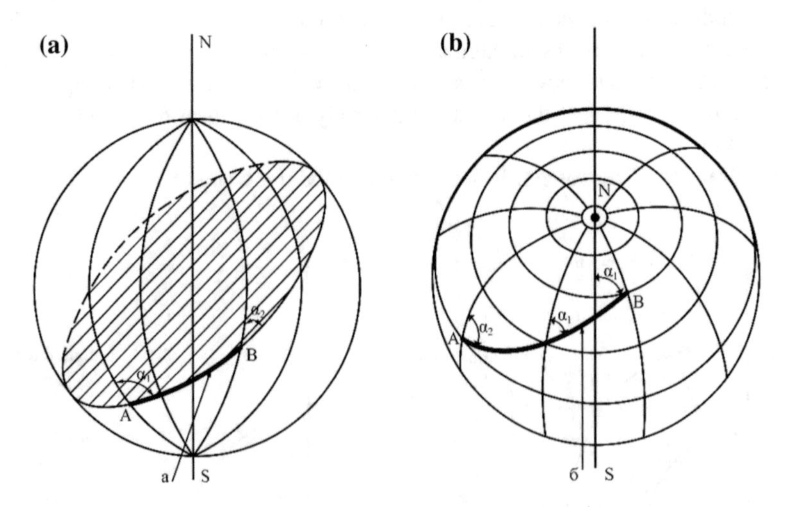

Fig. 1.2 Course lines of aircraft: **a** great circle, **b** rhumb line

1.3 Coordinate Systems

Local coordinate systems are used to calculate the position of the aircraft when flying over short distances. In this case, the curvature of the Earth's surface is neglected. Such calculations are carried out during takeoff and landing, in determining the waypoints for local airlines, etc. Figure 1.3 shows Cartesian, rectangular, spherical, and cylindrical coordinate systems used in such cases.

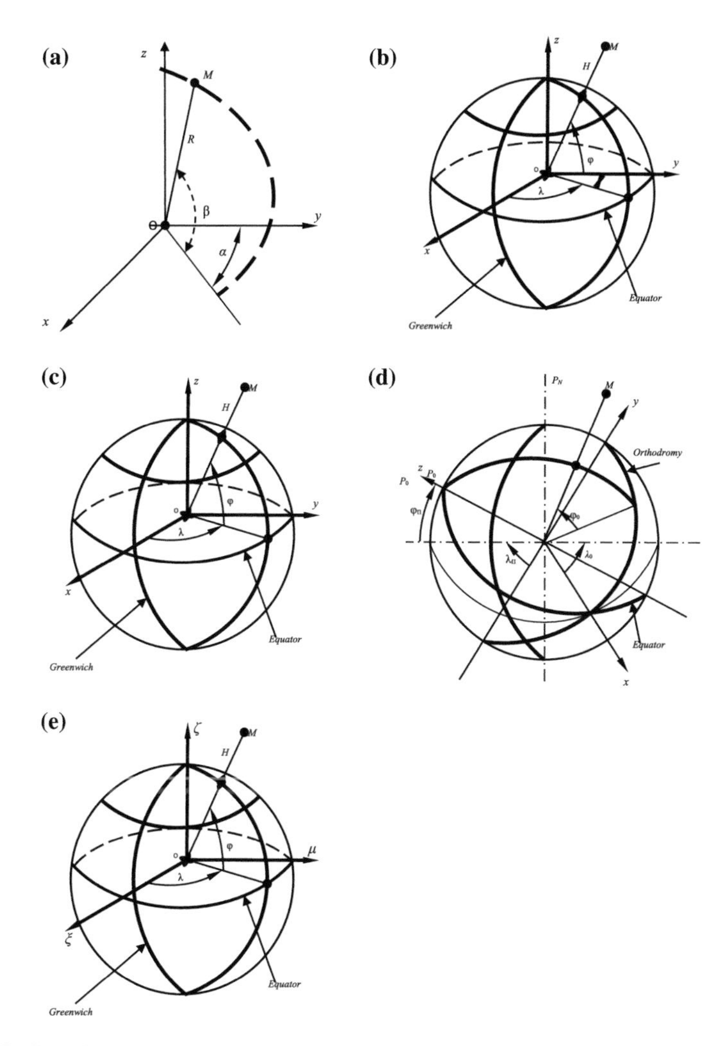

Fig. 1.3 Coordinate systems, **a** spherical, **b** geographical, **c** geocentric, **d** great circular, **e** geocentric inertial

Global coordinate systems are usually rigidly bounded to the Earth, covering the Earth's surface and near-Earth space. Global coordinate systems include geographic, geocentric, great circles, and Earth-centered inertial systems.

The equator is used as the reference plane in the geographic coordinate system. The coordinates are λ and φ (longitude and latitude respectively), relative height H.

In the geocentric coordinate system, the latitude is calculated between the equator plane and the radius vector of the object location point. In this coordinate system, the Earth is represented as a full sphere.

The great circles coordinate system is similar to the geocentric one and differs in that the plane of the large circle and the associated coordinate system are rotated relative to the Earth equator by some angle. The coordinates of the pole of great circle—$P_o(\lambda_p, \varphi_p)$, great circles coordinates—λ_o, φ_o H, R (radius vector of the M point).

Geocentric inertial coordinate system is used to determine the coordinates in the near-Earth space. The basic plane of reference in this system is the plane of the celestial equator. The reference point is aligned with the center of the Earth. The axes are directed to certain points of the celestial sphere (do not rotate with the Earth). The axis is directed to the point of the spring equinox (in the Aries constellation) and coincides with the line of the crossing of the planes of the equator and the Ecliptic, the axis ζ coincides with the axis of the Earth, the axis μ is perpendicular to the axes ξ and ζ. Coordinates in the system: λ—right ascension, γ—declination, R—length of the radius vector.

Let us consider the coordinate system for navigation of marine objects separately.

Using a variety of coordinate systems is not always convenient, especially in complex navigation systems, where information comes from different measuring instruments.

Geographical coordinate system

In this system, the coordinate axes are the Earth's equator and one of the meridians, conventionally taken for the initial one; coordinate lines are a meridian and a parallel passing through a given point, and the coordinates are geographic latitude and longitude.

Geographical latitude is the angle between the normal to the surface of the Earth ellipsoid at a given point and the plane of the equator (Fig. 1.4). A meridian arc from the equator to the parallel of a given point measures this angle. Geographic latitude is denoted by the letter φ. Latitudes are counted from the equator in both directions, i.e., North and South, from 0 to 90°. The northern latitudes are considered to be positive (sign +), they are denoted by the letter N; southern latitudes are considered to be negative (sign −), they are denoted by the letter S.

To count the longitude value, one of the meridians is taken as the initial or zero one, and the position of the other meridian, including the one passing through the place of the given M point, is determined relative to the initial meridian.

Geographical longitude is the dihedral angle between the plane of the initial meridian and the plane of the meridian passing through the given point. This dihedral angle is measured by a spherical angle at a pole between the specified meridians

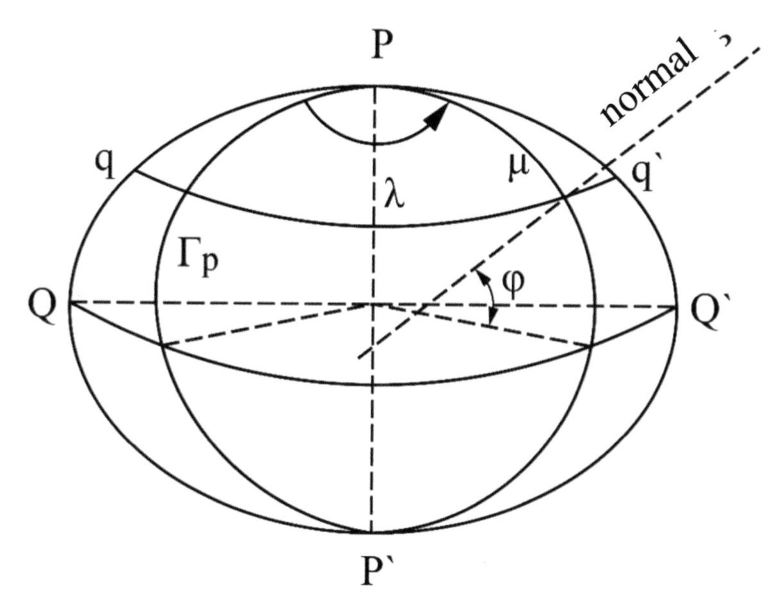

Fig. 1.4 Geographical coordinate system

or a numerically equal arc of the equator between the same meridians. Nowadays, Greenwich Meridian, passing through the former place of the Greenwich Observatory (near London) is determined as initial meridian under an international agreement of 1984. Geographical longitude is denoted by the letter λ.

Geographical longitudes are counted to the East and West of the initial meridian from 0 to 180°. Eastern longitude (0) is considered to be positive (sign "+"); Western longitude (*W*) is considered to be negative (sign "−").

Geodetic coordinate system

In this system, the position of a point on the surface of an ellipsoid is determined by the geodetic latitude *B* and the geodesic longitude *L*, which are geometrically treated equally with the corresponding coordinates of the geographic system. Geodetic coordinates are determined by geodetic measurements, for example, by reference to triangulation points reduced to the surface of the Earth ellipsoid. Geographical and geodetic coordinate systems refer to the mathematical law of the Earth—the ellipsoid of rotation.

Until 1946, the Bessel ellipsoid derived in 1841 was used as a reference ellipsoid. The dimensions of the Bessel ellipsoid are as follows: the large half-axis a = 6,377,397 m, the small half-axis b = 6,356,079 m, the polar compression α = 1:299.153.

Since 1946, the Krasovsky reference ellipsoid has been accepted as mandatory for all topo-geodesic works in the Russian Federation. The dimensions of the Krasovsky reference ellipsoid are as follows: a = 6,378,245 m, b = 6,356,863 m, α = 1:298.3.

Table 1.1 Reference ellipsoids

Ellipsoid	a (m)	$1/\alpha$
Airy	6,377,563.396	299.3249646
Australian National	6,378,160	298.25
Bessel	6,377,397.155	299.1528128
Clarke 1866	6,378,206.4	294.9786982
Clarke 1880	6,378,249.145	293.465
Everest	6,377,276.345	300.8017
Fischer 1960	6,378,155	298.3
Fischer 1968	6,378,150	298.3
GRS-80	6,378,137	298.2572221
Helmert	6,378,200	298.3
International (Hay-Ford)	6,378,388	297
Krassowski (1940)	6,378,245	298.3
NWL-9D = WGC-66	6,378,145	298.25
South American 1969	6,378,160	298.25
WGC-72	6,378,135	298.26

Table 1.1 shows the names and dimensions of some reference ellipsoids, used in different countries.

System of plane rectangular coordinates

This system, similar to the Cartesian coordinate system known from analytical geometry, is the most simple for practical use. Plane rectangular coordinates are called right-angle coordinates X and Y of the point depicted on the plane, which is a complex function of its geographical coordinates. Since the system of plane rectangular coordinates with a single reference for the entire Earth surface causes large distortions in the dimensions of the figures depicted on the plane, the entire Earth surface is divided into parts or coordinate zones.

In Russia, the zones with a length of 6° along the longitude are set. Thus, the entire surface of the Earth is divided into 60 six-degree zones; zones are counted to the East of the Greenwich Meridian (Fig. 1.5) from 1th to 60th number. The reference point of plane rectangular coordinates is the point of crossing of the equator and the middle meridian of each zone, called the axial meridian. The axial meridian of the zone and the equator are represented on the plane by mutually perpendicular straight lines, which are taken, respectively, as the X abscissa axis and the Y ordinate axis.

Plane rectangular coordinates X and Y are expressed in linear units. When the point is located to the North of the equator, then abscissa X is considered positive; for a point located South of the equator, abscissa X is negative. Ordinate Y is considered to be positive if the point is located to the East of the axial meridian of the zone, and negative if the point is located to the West of it. In order to avoid negative ordinate values, the axial meridian of each zone is given a value of 500 km. In addition, the zone number is placed before the ordinate value.

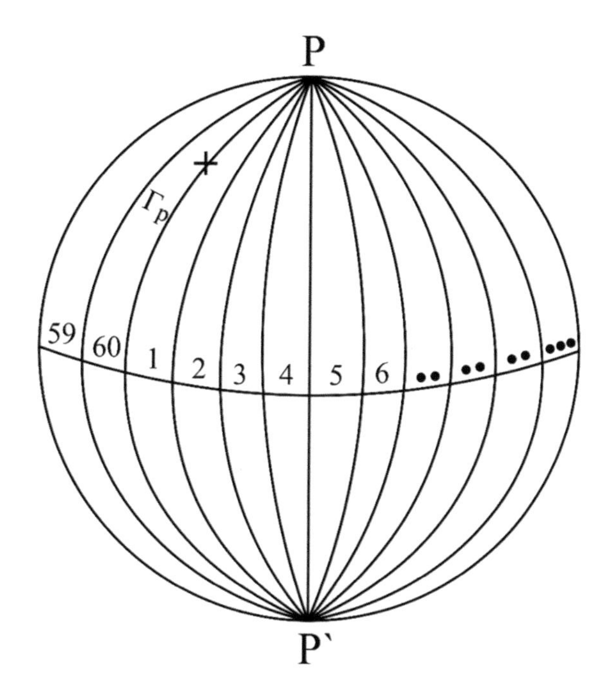

Fig. 1.5 Zones for system of plane rectangular coordinates

The formula for calculating the zone number for a known longitude is as follows:

$$n = \frac{\lambda}{6} + 1. \tag{1.5}$$

The coordinates of the object in a given zone are determined by using formula (1.5).

References

1. Bykov VI, Nikitenko YI (1976) Marine radio navigation devices. M. Transport, 399 pp
2. Yarlykov MS (1995) Statistical theory of radio navigation. M. Radio and Communication, 344 pp
3. Shatrakov YG (1999) Estimation of location error in the PI of the difference-ranging system and their dependence on the hyperbolic position lines. Col. Ed. Issue 1. VIMI, C58–95
4. Olianyuk PV et al (1995) Radio navigation systems of the ultra-long-wavelength range. M. Radio and Communication, 264 pp
5. Shatrakov YG (1990) Radio navigation systems of mobile objects. M. MRP, 92 pp
6. Shebshaevich VS et al (1989) Differential mode of network satellite RNS. Foreign Radio Electron 1:C5–32
7. Shebshaevich VS (1992) Network satellite RNS. M. Radio and Communication, 280 pp

Chapter 2
The Concept of Radio Navigation Systems

2.1 Features of Radio Waves Propagation

There are two areas in the surrounding atmosphere of the globe that affect the propagation of radio waves: the troposphere and the ionosphere [1, 2]. The troposphere is the surface area of the atmosphere, spreading up to a height of 10–15 km. The troposphere is heterogeneous both in the vertical direction and along the Earth's surface, in addition, its electrical parameters change with changing of meteorological conditions. The troposphere influences the propagation of radio waves and ensures the propagation of the so-called tropospheric waves. The propagation of tropospheric waves is associated with refraction in an inhomogeneous medium, as well as with the scattering and reflection of radio waves from inhomogeneities.

The ionosphere is the area of the atmosphere that starts at a height of 50–80 km and spreads approximately up to 10,000 km above the Earth's surface. In this area, the gas density is very low and the gas is ionized. The number of free electrons in the 1 cm^3 of the ionosphere is approximately 10^3–10^6. The presence of free electrons significantly affects the electrical behavior of the gas and makes it possible to reflect radio waves from the ionosphere. By successive reflection from the ionosphere, radio waves propagate over very long distances.

Outside the ionosphere, the gas density and electron density decrease, and conditions of radio wave propagation here are close to the conditions of propagation in vacuum space.

Radio navigation systems are designed using all frequency bands: extra-long waves (100,000–10,000 m), long waves (10,000–1000 m), medium waves (1000–100 m), meter waves (10–1 m), decimeter waves (1–0.01 m), and centimeter waves (0.01–0.001 m). The range of the RNS using meter, decimeter, and centimeter waves is limited by the direct visibility between the ground station and the moving object. And super long waves as well as long waves extend over long distances and can bend around the globe several times due to the properties of the surrounding global atmosphere.

Sauta O.I. et al., *Principles of Radio Navigation for Ground and Ship-Based Aircrafts*,
Springer Aerospace Technology, https://doi.org/10.1007/978-981-13-8293-2_2

There is an interfering spatial signal reflected from the ionospheric layer on the medium waves at the receiving point at the exact time, in addition to the useful surface signal. It is necessary to expand the spectrum of emitted signals to exclude it.

2.2 Classification of Radio Navigation Systems

Methods and means of radio navigation are classified according to the following basic principles [1]:

by the type of the measured navigation parameter (type of geometric quantity) or its time derivative);
by the type of radio technical measurements (depending on the type of the measured radio signal parameter used to determine the navigation parameter);
by appointment;
by operating range.

Methods and means of radio navigation are divided by the first parameter into goniometrical (tracking), range-measuring, differential range-measuring, linear (radial) and angular velocities measures, and combined, allowing to determine the various navigation options jointly.

The parameters of electromagnetic oscillations delivering navigation information are considered while dividing the navigation methods and means by the type of measurements. In this case, there are: phase, amplitude, frequency and time radio navigation systems (RNS).

RNS are divided by appointment:

- long-range navigation systems,
- short-range navigation systems,
- instrument landing, determining coordinates to prevent collisions;
- prevention of collisions,
- ground speed measuring.

RNS are divided by operating range:

- global, i.e., with unlimited operating range, allowing to determine the place of the object in any point of the globe or in near-Earth space
- long-range navigation, for objects at a distance of up to 3000 km from the radio navigation points (RNT), relative to which the spatiotemporal coordinates of objects are determined;
- short-range navigation, for objects at a distance of up to 400 km from the RNT.

In addition, radio navigation methods and means can be divided by other parameters:

- by the method of the coordinates determining—range-measuring, differential range-measuring, retrodirective range-measuring, range-measuring with very stable oscillators using, phase, goniometrical, differential, Doppler;
- by emission nature (with continuous and pulse emission);
- by autonomy level (autonomous and nonautonomous);
- by automatization level (automatic, semiautomatic, manual);
- by indication method (visual and audio indication).

The classification of radio navigation systems is also carried out by the method of signal measuring, the range of waves used, the place of equipment installation, the electromagnetic field polarization, etc.

2.3 Methods for Determining the Location of Moving Objects

A complex of radio technical equipment designed to determine the location of a moving object is called a radio navigation system [1–5]. Nowadays, there is a large variety of such systems. They include radio beacons, satellites, emitting radio navigation signals, control points, and equipment by means of which they receive radio navigation information, process it, and provide coordinates and other information to the executive bodies of moving objects. In practice, until recent years, there was a tendency to build equipment, through which the signal from beacons was received and processed with the use of universal computers. However, the equipment weight and size increased due to the excessiveness in the computing complex.

Development of microprocessors and miniature storage devices in recent years allowed to use the computers in the apparatus of such functionality, which excludes all typical computer excessiveness, the apparatus become compact, with minimal weight. This allows to expand the scope of application of radio navigation systems up to the personal use of indicators by individual groups of people, on small objects, including unmanned moving objects.

Let us consider the methods of determining the coordinates in radio navigation systems.

Range-measuring method. Is that signal is emitted from a moving object or from a beacon. The surface of position, in this case, is a sphere with a radius equal to the range of the system. The center of the sphere coincides with the location of the transmitting (receiving) device. In the Cartesian coordinate system, the surface of the position equation is written as

$$R^2 = x^2 + y^2 + z^2. \tag{2.1}$$

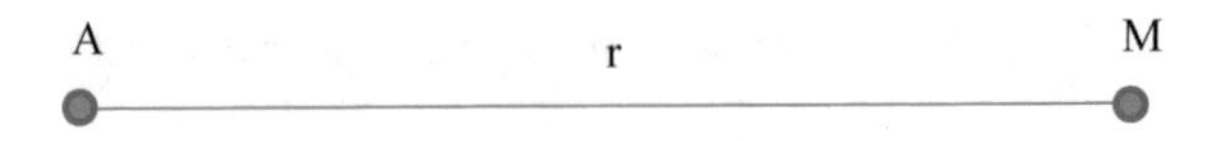

Fig. 2.1 Range-measuring method of determining the coordinates

Range-measuring method with very stable oscillators is applied to the phase systems. At point A (Fig. 2.1), the antenna of the transmitting device emitting a highly stable signal with frequency ω is placed. At point M, on the distance of $AM = r$ from point A, receiver in which the phase difference between the received signal and the signal of the reference generator having the same frequency ω is measured is placed. In this case, the measurement of the phase difference is equivalent to the measurement of the difference between the clock readings located at points M and the readings of the clock A transmitted by the radio signal. When the clocks are synchronous, then the time difference is $\tau = r/c$, where c—radio signal propagation speed. This time difference corresponds to the signal phase difference $\psi = \omega r/c$.

Inaccurate initial φ_o phase synchronization causes an error in determining the range

$$\sigma r = \frac{\delta\Psi}{k}[\text{rad}] = \frac{\lambda}{100}\delta\Psi[c.h.] = \frac{\lambda}{360}\delta\Psi\left[\text{deg}\right]. \tag{2.2}$$

Phase in radio navigation is counted in whole and hundredths of the phase cycle (c.h.), in the radio waves propagation theory in radians and degrees:

$$1\,c.h. = 3.6^\circ \approx 0.0628\,\text{rad};\ 1^\circ = 0.278\,c.h \approx 0.0174\,\text{rad};\ 1\,\text{rad} \approx 57.3 \approx 15.9\,c.h.$$

Frequency difference of generators δf leads to the error, depending on T time, passed since synchronization moment

$$\delta r_t = \lambda\delta\Psi_T = \lambda\int_o^T \delta\cdot f dt. \tag{2.3}$$

When the frequency difference δf during the T time can be adopted as permanent, then

$$\delta r_t = \lambda\delta f T = \lambda f(\delta f/f)T = c(\delta f/f)T. \tag{2.4}$$

When it is necessary to determine the coordinate with the accuracy of $\delta r_t = 1000$, 100, 10 m 5 h after synchronization, then we shall have relative generator instability less than, respectively,

$$\frac{\delta f}{f} = \frac{\delta r_t}{cT} = \begin{Bmatrix} \frac{1}{3 \times 10^5 1.8 \times 10^4} \approx 1.85 \times 10^{-10} \\ \frac{0.1}{3 \times 10^5 1.8 \times 10^4} \approx 1.85 \times 10^{-11} \\ \frac{0.01}{3 \times 10^5 1.8 \times 10^4} \approx 1.85 \times 10^{-12} \end{Bmatrix}. \tag{2.5}$$

Retrodirective range-measuring method is explained in the following. Signal from the M point at the t_o moment is transmitted to the A point, where clocks synchronization is conducted (Fig. 2.1) by received signal, i.e., at the $t_o + \tau$ moment of time (τ—propagation time). Then, at a point in time with a known delay t_k, often referred to as a base or code, a signal containing information on the clock reading at point a is transmitted. This signal comes to the point M at the $t_o + \tau + t_k + \tau$ time. Thus, at point M, it is known that, given the time of propagation, the difference in the clock reading is $\Delta t = 2\tau + t_k$. The corresponding phase shift value will be equal to

$$\Delta\Psi = \omega\Delta t = \omega(2\tau + t_k) \tag{2.6}$$

Code delay t_k is within a few microseconds to seconds depending on the technical principles of the system. Obviously, the requirement for the accuracy of the clock, arising from the expressions, becomes easily feasible in this case. In particular, to ensure $\delta r_t = 3$ m, the frequency stability must be 10^{-8} for 1 s, which is easy to practically implement

$$\frac{\delta f}{f} = \frac{\delta r_t}{cT} = \frac{3}{3 \times 10^8 \cdot 1} = 1 \times 10^{-8}. \tag{2.7}$$

Differential range-measuring method. The distance difference between the moving object and the two radio navigation stations is

$$\Delta R = R_1 - R_2 = 2\text{d} \tag{2.8}$$

The surface of position, which corresponds to the distance difference, is a hyperboloid of rotation around the axis, coinciding with the basis of the radio navigation stations. Figure 2.2 shows the specified surface of the position.

The surface of the position equation is determined by the formula:

$$\frac{x^2}{a^2} + \frac{y^2}{b^2} + \frac{z^2}{c^2} = 1, \tag{2.9}$$

where a, b—semi-transverse axis; c—imaginary semi-axis.

Differential range-measuring method does not require absolute synchronization and does not need the presence of a transmitting device on the moving object for phase measurements. By the signal emitted at point A (Fig. 2.3) at time t_o, clocks are

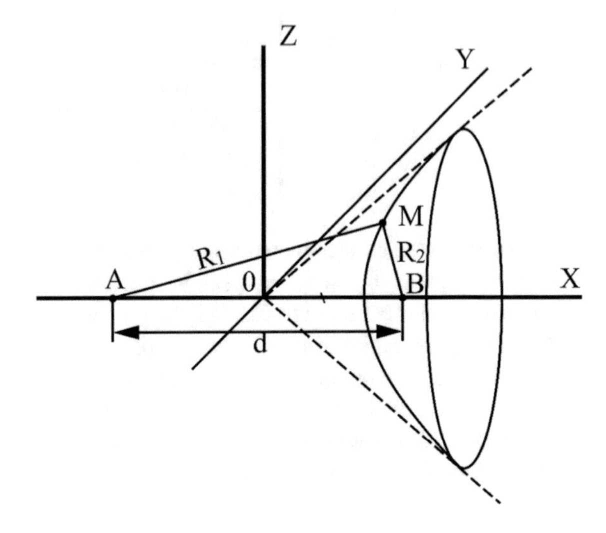

Fig. 2.2 Differential range-measuring method for coordinates determining—the surface of the position

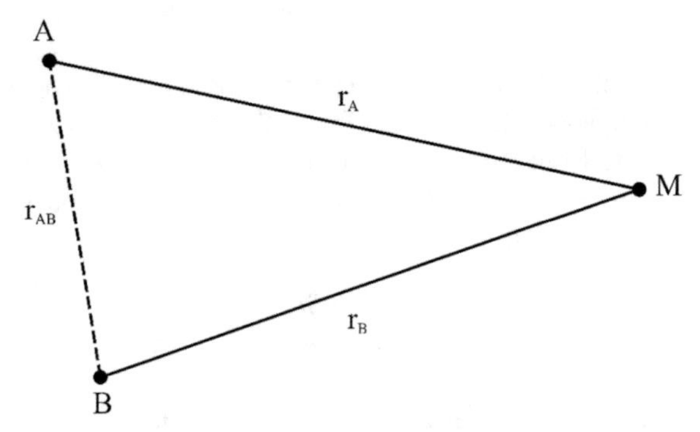

Fig. 2.3 Differential range-measuring method for coordinates determining (geometric equations for estimating the delay of arrival of signals from two stations)

synchronized at point B at $t_0 + \tau_{AB}$, where $\tau_{AB} = r_{AB}/c$. Then, a message is sent with a delay, which is received at the object at a time

$$t_B = t_0 + \tau_{AB} + t_k + \tau_B \tag{2.10}$$

Here $\tau_B = \frac{r_B}{c}$, $\tau_{AB} = \frac{r_{AB}}{c}$, c—is light speed.

The message sent from point A is received at point M at the time

$$t_A = t_0 + \tau_A.$$

Here $\tau_A = r_A/c$.

The difference between the time of receiving messages, equal to

$$\Delta t = t_B - t_A = \tau_{AB} + t_k + \tau_B - \tau_A \tag{2.11}$$

does not depend on the time t_0, that is, accurate synchronization is not required. The phase difference corresponding to Δt:

$$\Psi = \omega \Delta t = \omega(\tau_{AB} + t_k) + k(r_A - r_B) \tag{2.12}$$

The first term is a known constant because t_k and r_{AB} are known. So Ψ is defined by range difference $r_B - r_A$.

Reversed differential range-measuring method. This method is used in the case where the transmitting device can be installed on a moving object, and the determination of its coordinates must be carried out at a fixed point. At the time $t_{0,}$ the signal is transmitted from point M (Fig. 2.3). Station at the point B operating as in differential range-measuring method, passing the message of its clocks readings synchronized by a signal from point M. In this case, the difference between the time of receiving messages at point A, coming from points M and B, and the corresponding phase difference will be expressed by the same formulas (2.11) and (2.12).

Another option of differential range-measuring method is possible when the signal is transmitted at the time t_A from the point A. This signal is received at point M at time $t_A + \tau_A$ and reemitted at the time $t_M = t_A + \tau_A + t_{K(M)}$.

Interval $t_{K(M)}$—code delay of station K. The signal transmitted from point M is used to determine the distance difference $r_B - r_A$ at point A.

By measuring the difference between the arrival of signals t_M at the point M from the points A and B, receiving

$$t_{BA} = \tau_{AB} - t_{K(M)} + \tau_B - \tau_A = \tau_{AB} + t_{K(M)} + \frac{r_B - r_A}{c} \tag{2.13}$$

Differential method. In cases where stations synchronization is hindered, a differential or compensatory method is used (Fig. 2.4).

Suppose at the time t_A station A transmits a signal. Regardless of this, at the time t_B station B emits the signal. These signals will be received by the control point, located at point N, respectively, at the time $t_A + \frac{r'_A}{c}$ and $t_B + \frac{r'_B}{c}$, after that, clocks are synchronized at this point. Then after the t_k time their readings are transmitted and will be received at the point M in time $t_k + \frac{r_{MN}}{c}$ after receiving relevant signals in the point M, i.e., at the time

$$t_{MA(N)} = t_A + r'_A/c + t_k + \tau_{MN}. \tag{2.14}$$

$$t_{MB(N)} = t_B + r'_B/c + t_k + \tau_{MN}. \tag{2.15}$$

Here τ_{MN}—signal propagation time from the point M to N. Previously, the point received signals transmitted from points A and B at the time

$$t_{MA} = t_A + r_A/c. \tag{2.16}$$

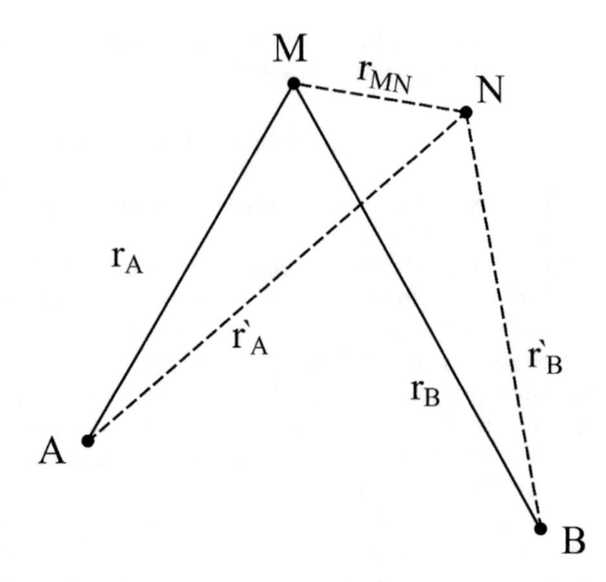

Fig. 2.4 Differential method for coordinates determining

$$t_{MB} = t_B + r_B/c. \quad (2.17)$$

After the differences measuring

$$\Delta t_A = t_{MA(N)} - t_{MA} \quad (2.18)$$

$$\Delta t_B = t_{MB(N)} - t_{MB} \quad (2.19)$$

receiving

$$\Delta t_A = r'_A/c + t_k + \tau_{MN} - r_A/c \quad (2.20)$$

$$\Delta t_B = r'_B/c + t_k + \tau_{MN} - r_B/c \quad (2.21)$$

When the frequencies of A and B stations are approximately the same, i.e., $\omega_A \approx \omega_B = \omega$, $\Delta\omega = \omega_A - \omega_B \ll \omega$, then it is possible to determine phase difference

$$\Psi = \omega_A \Delta t_A - \omega_B \Delta t_B \approx \omega(r_B - r_A)/c - \omega\left(r'_B - r'_A\right)/c + \Delta\omega(t_k + \tau_{MN}) \quad (2.22)$$

It follows from the formula (2.22) that this method does not impose high requirements for the exact knowing of the frequency $\Delta\omega$ and time $t_k + \tau_{MN}$ difference.

Range-measurement goniometric method. In this case range, Rand angular points α and β are determined. When using a spherical coordinate system (R, α, β), the angle in the horizontal plane is determined by the coordinate α. The surface of

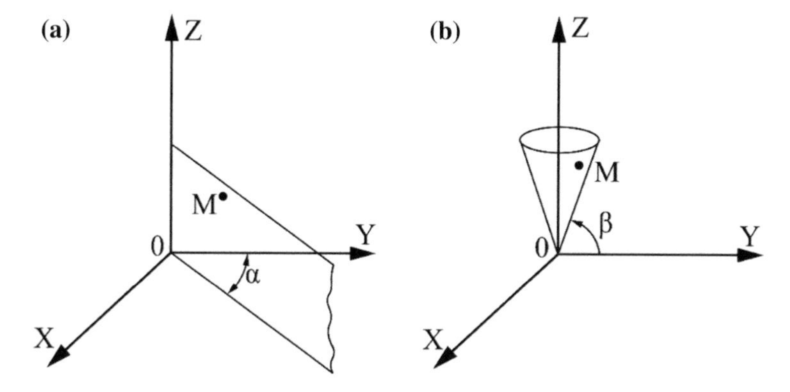

Fig. 2.5 The surface of position for determining the angle in horizontal (**a**) and vertical (**b**) planes for range-measurement goniometric method for coordinates determining

position will be vertical plane, located at this angle. The reference direction is usually North. The surface equation can be written as

$$\alpha = \arctan\left(\frac{x}{y}\right) \tag{2.23}$$

The surface of position for the range-measurement goniometric method for coordinates determining is shown in Fig. 2.5a.

The β coordinate determines the vertical plane angle. The surface of position is a cone obtained by rotating a straight line around a vertical axis. Figure 2.5b shows this surface of position.

The equation of the surface of position for determining the angle in the vertical plane can be written as

$$\beta = \arctan\left(\frac{z}{\left(x^2 + y^2\right)^{1/2}}\right). \tag{2.24}$$

Range-measurement methods are measuring the distances from the moving object to several radio navigation stations with known coordinates. When crossing the surface of the position of the sphere with the plane, we obtain a line of position in the form of a circle. It is necessary to determine three ranges since the circle does not allow determining the coordinates unequivocal.

Figure 2.6a shows the location of the moving object as the point M, which is determined as the point of three circles crossing (lines of the sphere with the plane crossing).

Differential range-measuring methods determines position line as hyperbolic curve. The object at point M determines two differences of the distances

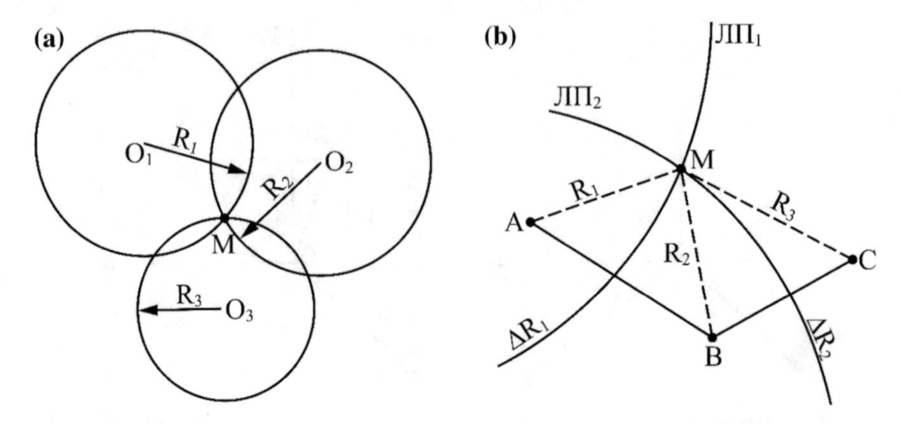

Fig. 2.6 Determining moving object position by three ranges (**a**) and by differential range-measuring method (**b**)

Fig. 2.7 Bearing method for coordinates determining

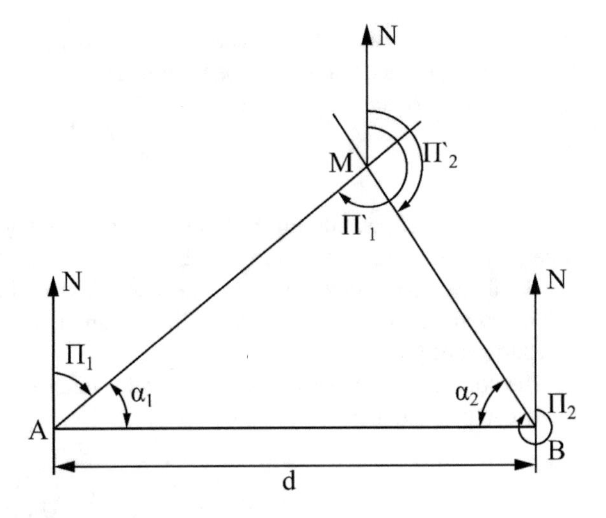

$$\Delta R_1 = R_1 - R_2, \Delta R_2 = R_2 - R_3. \quad (2.25)$$

The found value ΔR_1 corresponds to the position of line-hyperbolic curve LP_1, and ΔR_2 corresponds to LP_2 (Fig. 2.6b). Object location determined as the point of two line positions crossing. Stations A, B, C are located in such way to eliminate the double intersection of hyperbolic curves in the stations operation area.

Bearing (goniometric) method. Here are two bearings P_1 and P_2 to the object located at the point M (Fig. 2.7).

Point M coordinates can be found by ABC triangle solving by known basis d and α_1 and α_2 angles. The place of M on moving object can be also determined in a such way by measuring P_1' and P_2' bearings. Knowing P_1' and P_2', P_1 and P_2 can be found (they are called reverse bearings in that case):

$$P_1 = P_1' - 180^\circ. \tag{2.26}$$

$$P_2 = P_2' + 180^\circ. \tag{2.27}$$

Triangle ABM is solved like in the first case.

The location of the moving object in space is determined by the measurements of three parameters in the radio navigation station coordinate system. Real equipment determines coordinates with errors, so the location of the object is also determined with an error.

Position error is the distance between the measured and true object position.

Surface of position error. As known, each value of the measured coordinate in the radio navigation system corresponds to a certain surface of position. When the navigation parameter is measured with an error, the corresponding position surface will be determined with an error. The surface of position error at the location point of the object is the normal distance between the true and the measured surfaces of position. Let us take a rectangular coordinate system x, y, z for consideration. Then the measured by the radio navigation system coordinates can be expressed, in general, as

$$P = f(x, y, z). \tag{2.28}$$

When small increments P are set, then there will be a transition from one surface of position in the radio navigation system coordinate system to the neighboring, nearby surface of position. Therefore, Eq. (2.28) defines the surface of some level in a continuous rock field. The gradient P module will determine the rate of change of the second field in the direction of the normal to the surface. In a rectangular coordinate system

$$grad P = i\frac{\delta P}{\delta x} + j\frac{\delta P}{\delta y} + k\frac{\delta P}{\delta z}, \tag{2.29}$$

respectively,

$$|grad P| = \left[\left(i\frac{\delta P}{\delta x}\right)^2 + \left(j\frac{\delta P}{\delta y}\right)^2 + \left(k\frac{\delta P}{\delta z}\right)^2\right]^{1/2}. \tag{2.30}$$

It can be written by changing the differentials to finite increments

$$\Delta n = \frac{\Delta P}{|grad P|} \tag{2.31}$$

When ΔP is an error of P parameter measurement, then Δn defines position error. Suppressing intermediate conversions, writing

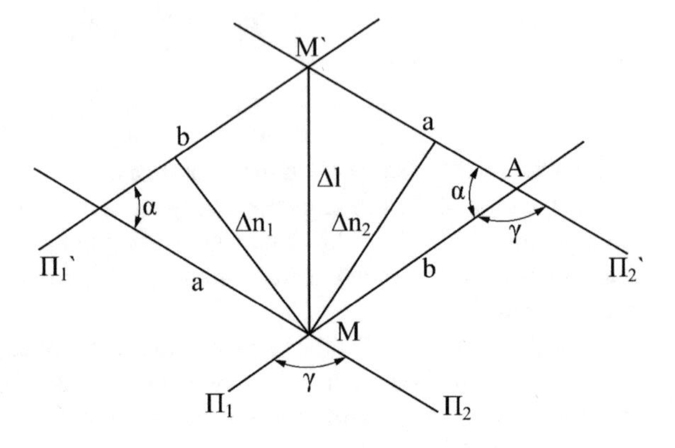

Fig. 2.8 Line of position error

$$\Delta\sigma_n^2 = \frac{\sigma_P^2}{|gradP|^2} \tag{2.32}$$

When the error variance of parameter σ_P^2 measuring is known, then it is possible to determine the surface of position error variance.

Surface of position error will be different in the different points of space because of the variable $|gradP|$ over a constant σ_n parameter error.

Line of position error. The line of position is the result of the two surfaces of position crossing. Line of position error is the normal distance between true and measured lines of position. Point M at Fig. 2.8 is the trace of the true line of position on the drawing plane. P_1 and P_2 are the sections of the two surfaces of position by drawing a plane.

Measured surfaces of position P_1' and P_2' crossing form measured line of position, and M' is the trace of it. MM' distance, which is equal to Δl, is the line of position error Δl value can be found from AMM' triangle

$$\Delta l = \left(a^2 + b^2 - 2\text{abcos}(\alpha)\right)^{1/2} \tag{2.33}$$

a, b, α can be expressed through the surfaces of position $\Delta n_1, \Delta n_2, \gamma$ errors

$$a = \frac{\Delta n_1}{\sin(\alpha)}; b = \frac{\Delta n_2}{\sin(\alpha)}; \alpha = \pi - \gamma \tag{2.34}$$

Then

$$\Delta l = \frac{1}{\sin^2(\gamma)}\left(\Delta n_1^2 + \Delta n_2^2 - 2\Delta n_1 \Delta n_2 \cos(\gamma)\right)^{1/2} \tag{2.35}$$

Going over to mean squares, receiving expression for error variance

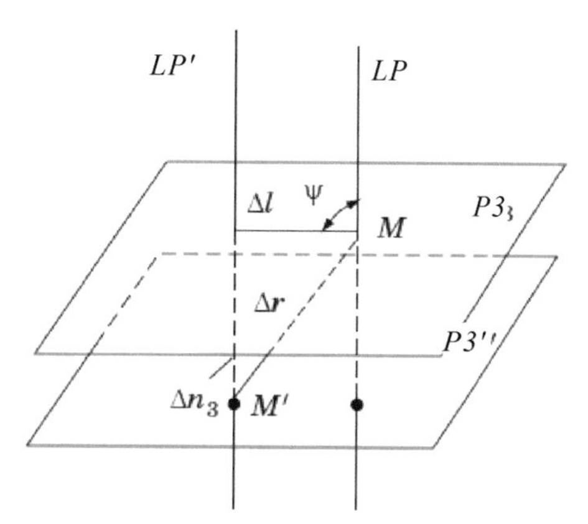

Fig. 2.9 Position in space determining error

$$\sigma_l^2 = \frac{1}{\sin^2(\gamma)}\left(\sigma_{n1}^2 + \sigma_{n2}^2 - 2\mathrm{M}_{1,2}\cos(\gamma)\right)^{1/2}, \quad (2.36)$$

where $M_{1,2} = \overline{\Delta n_1 \Delta n_2}$—is the errors mean product

With underlying $\gamma = \frac{\pi}{2}$ error is maximal and equal to

$$\sigma_l^2 = \left(\sigma_{n1}^2 + \sigma_{n2}^2\right)^{1/2}. \quad (2.37)$$

Position in space error. Object position in space is determined as the point of line of position and the third surface of position crossing. Figure 2.9 shows LP, P_3, and M—true line of position, surface of position, and object position.

LP^1, P_3^1, M^1—measured, ψ—LP and P_3 crossing angle, Δl, Δn_3, Δr—line of surface and position errors. Suppose that Δl, Δn_3, ψ are specified.

Calculation formulas become complicated, and for the case $\psi = \frac{\pi}{2}$

$$\Delta r^2 = \Delta l^2 + \Delta n_3^2 \quad (2.38)$$

$$\Delta\sigma_r^2 = \Delta\sigma_l^2 + \Delta\sigma_{n3}^2 \quad (2.39)$$

Line of position determining on surface error. Line of position on surface error xOy can be received as special case from (2.31) and (2.32) formulas, if the space is taken as two dimensional. In this case, the measured parameter corresponds to the curve on the plane (line of position) $P = f(x, y)$, and line of position errors will be defined by the following formulas:

$$\Delta l = \frac{\Delta P}{|grad P|} \quad (2.40)$$

$$\Delta\sigma_l = \frac{\Delta\sigma_P}{|grad P|} \quad (2.41)$$

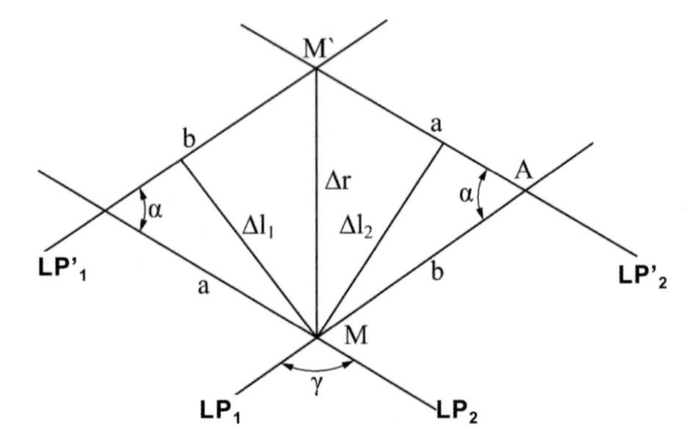

Fig. 2.10 Error of position on plane determining

where $|grad P| = \left[\left(\frac{\delta P}{\delta x}\right)^2 + \left(\frac{\delta P}{\delta y}\right)^2\right]^{1/2}$.

Error Δr depends here from the line of position Δl_1, Δl_2, γ errors. Figure 2.10 shows the necessary geometrical relations.

When these lines of position are assumed as specified, then

$$\Delta r = \frac{1}{\sin^2(\gamma)}\left(\Delta l_1^2 + \Delta l_2^2 - 2\Delta l_1 \Delta l_2 \cos(\gamma)\right)^{1/2} \quad (2.42)$$

$$\Delta \sigma_r = \frac{1}{\sin^2(\gamma)}\left(\sigma_{l1}^2 + \sigma_{l2}^2 - 2\rho\sigma_{l1}\sigma_{l2}\cos(\gamma)\right)^{1/2} \quad (2.43)$$

where ρ is the correlation coefficient.

2.4 System Working Area

The radio navigation system working area is limited by the amount of space within which the established accuracy of its operation is provided at a given ratio U_s / U_{sh} and at the input of the receivers, and the error in the location of the object determining with a given probability does not exceed the set value. The above area is always determined by the vector radius of the spatial figure, where the above conditions are met.

Working area for the ground-based systems very often repeats the shape of the directional diagram of the transmitting antenna, taking into account the nature of the receiving antenna gain ratio change on the moving object in its evolution. For sectoral radio navigation systems, which also include heading and elevation transmitters of instrumental landing, the shape of the emission pattern in the horizontal plane repeats the shape of the working area.

Range-measuring system. It consists of two stations located on the base ends. Assume that transmitters have the same accuracy σ_o. Then, while suppressing inter-

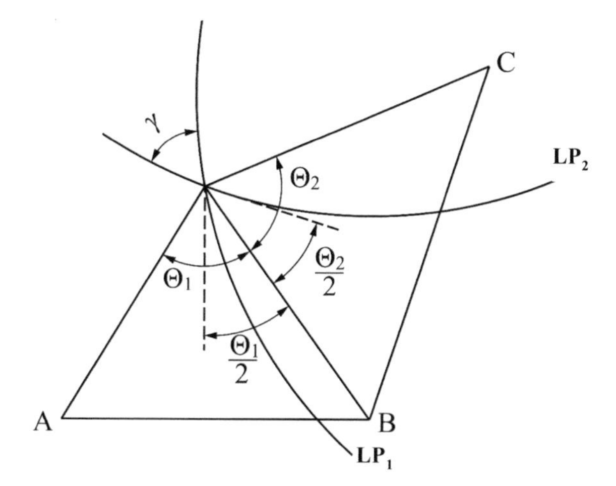

Fig. 2.11 Differential range-measuring radio navigation system working area calculation

mediate conversions, we can define the formula for system working area on the plane calculation

$$r_\sigma = \frac{\sigma_o}{\sqrt{2}\sin(\gamma)}(1 + \rho\cos(\gamma))^{1/2}, \tag{2.44}$$

where γ is the line of positions, drawn from the transmitter points of positions, crossing.

The curve of equal place error value will correspond to the constant angle of lines of position crossing. This curve will be a circle in which chord is the system base. The radius of the circle and the position of the center are determined from the formulas

$$R = \frac{d}{2}\cos(\gamma);\ y_u = \frac{d}{2}ctg(\gamma);\ x_u = 0 \tag{2.45}$$

Minimal place error will be when lines of position crossing at 90° at the circle, which diameter is equal to the system base

$$r_{\sigma\ min} = \frac{\sigma_0}{\sqrt{2}}. \tag{2.46}$$

Differential range-measuring system. It consists of three stations. Figure 2.11 shows geometrical relations. Place of the object is determined as the line of positions crossing for two AB and BC crossing.

The relationship between the angle of lines of position crosses γ to the angles Θ_1 and Θ_2 under which the base is viewed from point M, can be determined from the properties of hyperbolic curves. This property is that the tangent to the hyperbolic curve is the bisector of the inner angle between the radius vectors (i.e., AM and BM), drawn from the focus points (A and B). Thus,

$$\gamma = \frac{\Theta_1}{2} + \frac{\Theta_2}{2}. \tag{2.47}$$

Cross-correlation coefficient here is not equal to zero, because the measurements are depending (point B is common for both bases). Usually, the parameter measurement error (distances differences) for the two system bases are the same. Then a formula for the normalized position error can be written as

$$\sigma_{норм} = \frac{\sigma_r}{\sigma_p} = \frac{1}{2\sin\left(\frac{\Theta_1 + \Theta_1}{2}\right)} \left[\frac{1}{\sin^2\left(\frac{\Theta_1}{2}\right)} + \frac{1}{\sin^2\left(\frac{\Theta_2}{2}\right)} + 2\rho \frac{\cos\left(\frac{\Theta_1 + \Theta_2}{2}\right)}{\left|\sin\left(\frac{\Theta_1}{2}\right)\right| \left|\sin\left(\frac{\Theta_2}{2}\right)\right|} \right]^{1/2} \tag{2.48}$$

where σ_r is the position on the plane error;

σ_p is the distance difference (system parameter) measurement error;

Θ_1, Θ_2 are angles, under which bases are viewed from the lines of position crossing point;

ρ is the correlation coefficient.

Correspondence $\Theta_1 = f_1(x, y)$, $\Theta_2 = f_2(x, y)$ can be determined by specifying bases sizes and position in Cartesian rectangular coordinates (x, y). The result is an equation in the form of $f(x, y, \sigma_н, \rho) = 0$ Lines of equal errors, which are the limits of the working area with acceptable position errors $\sigma_r = \sigma_{норм}\sigma_p$ can be built by solving this equation.

2.5 Time Keeping in Radio Navigation Systems

Synchronizing the operation of the moving object onboard instruments and ground systems is a very difficult but extremely necessary task. Not only the accuracy of radio systems depends on the accuracy of synchronization , but also in some cases their construction principle. Thus, with accurate synchronization, it is possible to introduce request-free operation modes in some radio navigation and radar systems, which not only increase the tactical characteristics of moving objects but also create working conditions without interference.

Synchronization of radio technical equipment depends on the operation of time standard devices or systems, which are used on the moving object and ground equipment. The principle of building time standard onboard systems (synchronization devices) is based on the presence of autonomous timekeepers connected with the control system, ground-based synchronization equipment and ground-based timekeepers. The latter is due to the fact that in ground conditions it is easier to provide higher stability of the timekeeper, and it is possible to remove the accumulated time error in the onboard synchronization system through the control system. Depending on the requirements for the stability of time characteristics, quartz generators, quantum mechanical frequency standards, including atomic-beam ones, are used as

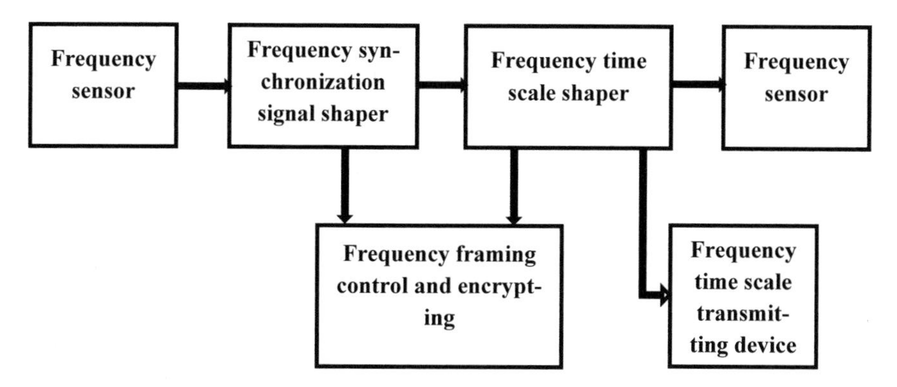

Fig. 2.12 Structural configuration of synchronization device

a timekeeper onboard of the moving object. Extensive studies have been carried out on the possibility of using devices on new physical principles of operation (hydrogen and laser standards) as onboard timekeepers. Figure 2.12 shows the block diagram of the onboard synchronization devices.

The formation of the synchro frequencies grid is carried out using digital dividers, conversion cells, or other computer elements. The generated synchro frequencies in the form of pulse synchro signals enter the radio technical systems on the moving object. Second marks (frequency signals in Hertz) are digitized in the timescale generator and in the form of a pulse, serial or parallel code is received by the onboard systems users of the moving object.

Considering the importance of information transmitted from the onboard timekeeper to the onboard systems, the connection between them is usually carried out on duplicated and even triplicated circuits for the purpose of the organization majority device for receiving time signals at the entrance of the onboard devices.

Digitization of marks is carried out within the accepted interval for the given system of standard time. The control of the onboard timekeeper operation is carried out by transmitting the 1 Hz mark and its digitization to the ground-based measuring points via the radio channel. The structure and method of transmission of this information depend on the construction of radio complexes.

Ship-based and ground-based timekeepers verification methods are divided into passive and active. With the passive time verification method, the ground-based point receives a signal transmitted from the moving object and records the value of the onboard timescale. The signal propagation time is calculated and determined based on the data of trajectory measurements. Parameters that characterize the state of the environment in the propagation path of the signal are also taken into account in that case. According to a series of measurements, statistical processing of signals is performed and the differences of ship-based and ground-based keepers timescales are determined.

With the active method of verification, the command-trajectory radio channel is used, in which the distance to the moving object is measured and the time difference of the scales in the system is determined. Refractive and other known errors are taken into account by means of calculations. The active method is more accurate, but requires additional equipment at ground-based verification points and onboard of the moving object. This reduces the autonomy of the moving object.

Increasing the accuracy of timescales storage in radio navigation systems can be achieved by

- increasing the stability of the onboard timekeeper reference generator;
- more frequent using of reconciliation and putting corrections to the onboard timekeeper scale;
- forecasting timescale and the correction range by calculated amendments in the intervals between the verification and correction operations.

There is also a distinction made between correction of the timescale of the onboard keeper by changing the lapse of time or the position of the timestamp of the keeper (phasing and correction), or by mathematical accounting of time information in the radio navigation signal (without control effect to the onboard keeper).

In the space navigation system, onboard of the spacecraft four timekeepers (atomic) were installed. The control subsystem consists of five tracking stations, main tracking station, as well as three stations for putting navigation and time information. Time information from each spacecraft is transmitted to two tracking stations at the same time, allowing the station scales to be synchronized with each other.

The equipment of the tracking stations includes two sets of ground-based timekeeper (one operating, one in the hot standby). They form the timescale of the navigation system. The main control station processes the measurement data received from the stations and makes the evaluation of the states of operation of onboard timekeepers (signal phases, signal frequencies, aging, etc.), ground-based keepers of time tracking stations (signal phases and frequencies) and the ephemeris of the spacecraft. The obtained results are regularly transmitted onboard of the spacecraft using the service information transfer stations. In addition, the correlation coefficients, forecasted by Kalman estimation of the scale shift of onboard timekeepers of each spacecraft are transmitted onboard of the spacecraft. As a reference timescale in the GPS system, the scale of the USA Naval Observatory was chosen. The error of the binding of the consumer system timescale on the USA Naval Observatory timescale shall not exceed 110 ns (I σ).

This total error is distributed by components as follows:

- ground-based timekeepers timescale error to the USA Naval Observatory timescale—not more than 90 ns (I σ);
- onboard timekeepers timescale error to the USA Naval Observatory timescale—not more than 97 ns (I σ);

Kalman estimation of the scale shift of onboard timekeepers error relative to the ground-based timekeepers of the tracking stations—not more than 7.5 s (I σ).

Characteristics of frequency standards in terms of its instability (Allan variance) are the following:

for on-board per $\tau_n = 1 сутки$ $2 \cdot 10^{-13}$ (I σ)

for ground-based per $\tau_n = 2ч$ $3 \cdot 10^{-14}$ (I σ)

To obtain information about the current time from the system, the consumer needs to determine the difference between its own timescale and the timescale of the systems ΔT_S. The carriers of the latter are, as it was shown above, the onboard timekeepers of spacecraft. The information is transmitted to the consumer from the spacecraft in the form of a second time mark corresponding to a certain phase of the code. The number of seconds within a day and the correction of ΔT_E for the onboard keeper shift relative to the time of the system is transmitted as the part of the information from the spacecraft

$$\Delta T_s = PR + \Delta T_E - R / C - \tau \tag{2.49}$$

where PR—pseudorange to the spacecraft;

ΔT_S—correction for the onboard timescale shift relative to the ground-based timekeeper scale;

R—geometric range to the given spacecraft;

C—light speed;

τ—additional delay of radio signal propagation in the ionosphere, troposphere, and consumer equipment

The consumer using the receiver automatically measures pseudorange PR, defines the ΔT_E value embedded in a message as a part of service information from the spacecraft. Then the value of the geometric range R is calculated using the ephemeris of the spacecraft, also transmitted in the service information. Further, τ is determined based on a priori known models of the ionosphere, troposphere, and data on the signal delay in the consumer equipment during processing. This procedure is performed several times to improve accuracy.

The variance of the timescale formation error caused by the frequency shift between the correction moments can be described by equations for rubidium and cesium standards, respectively,

$$\sigma^2(t) = 10^{-20}(t - t_k) + 1.44 \times 10^{-24}(t - t_k)^2, \tag{2.50}$$

$$\sigma^2(t) = 2.5 \times 10^{-21}(t - t_k) + 5.76 \times 10^{-26}(t - t_k)^2, \tag{2.51}$$

where t—is the current time; t_k—is the correction time.

The control and management subsystem in the system adjusts the timescale of the onboard keepers with such an interval that the onboard keeper mean square error does not exceed 12 ns.

Using the equipment operating at two frequencies f_1 and f_2 in the system, allows calculating the ionospheric delay and reducing the error to the few nanoseconds. In the troposphere, due to the use of models, it is possible to calculate the propagation delay with an accuracy of up to 1 ns.

Noise errors in the receiving signals in the system are a few nanoseconds due to the narrow band receiving.

References

1. Bykov VI, Nikitenko YI (1976) Marine radio navigation devices. M. Transport, 399 pp
2. Shatrakov YG, et al (1991) Marine RNS. M. Radio and Communication, 94 pp
3. Makarov GI, et al (1973) Propagation of an electromagnetic pulse above the earth surface. In: Problems of diffraction and propagation of radio waves. Leningrad State University, 95 pp
4. Shatrakov YG, et al (1986) Modern systems for short-range radio navigation of aircraft. M. Transport, 402 pp
5. Baburov VI, Ponomarenko BV (2005) Principles of integrated onboard avionics. SPb, "RDK-Print", 448 pp

Chapter 3
Short-Range Navigation Systems

3.1 Short-Range Radio Technical Navigation System

Domestic short-range radio technical navigation system (SRNS) has an azimuth channel and a range channel [1–5]. The azimuth channel on the moving object allows detection of the angle to the radio beacon. The operation principle is temporary. On the azimuth channel, the radio beacon antenna pattern has a two-lobe form in the horizontal plane and can rotate at a frequency of 100 rev/min.

The transmitter emits unmodulated or pulsed oscillations. The receiving device of the onboard equipment gates a pulse formed during emission of a moving object with a directional pattern. At the azimuth of the receiving point, this pulse, equal to A, is delayed by the time relative to the origin of the countdown, when the minimum of the radio beacon's directional pattern coincides with the northward direction of the meridian passing through the radio beacon installation point.

At this, the rotation frequency of the azimuth antenna of the radio beacon is equal to 100 rpm (600°). Time meters are used to allocate information about azimuth. A reference signal is used in order to fix on the moving object of the time reference. It is emitted by an undirected lighthouse antenna and contains two pulse sequences: one consists of 35 pulses, and the second consists of 36 pulses (for a period of one revolution of the directional azimuth antenna). At the moment when the minimum of the diagrams of the azimuth antenna passes through the North direction of the meridian, the pulses of the 35th and 36th sequences coincide.

This is considered to be the time reference point. When the minimum of the diagram of the azimuthal antenna passes through the moving object, the moment when the minimum of the diagram passes through moving object is fixed and the arrival time of the pulses of the 35th and 36th sequences is being compared. Further, the azimuth of the moving object is calculated relative to the lighthouse of the systems.

The range channel operates on the basis of the request–response principle and realizes an impulse method for determining the range. The range information is based on the interval of time between the moment of emission from the moving object of the request signal and the moment of receiving the response signal from the range-measuring lighthouse. This moment of time is determined as follows:

Sauta O.I. et al., *Principles of Radio Navigation for Ground and Ship-Based Aircrafts*,
Springer Aerospace Technology, https://doi.org/10.1007/978-981-13-8293-2_3

$$t = \frac{2D}{c} + t_{\text{del}} \tag{3.1}$$

where D is the range from the moving object to the lighthouse; c is the speed of light; t_{del} is the signal delay in the radio beacon.

The request signals for the range from the moving object consist of three pulses, and the response signals consist of two pulses. The pulse character of the ranging signals let the radio beacon work simultaneously with 100 moving objects.

The main parameters of a land-based lighthouse:

- the frequency range: 962–1000.5 MHz;
- range: 400 km (at altitude 10 km);
- measurement error of azimuth: 0.25° (2σ);
- range measurement error: 200 + 0.03% D (m) (2σ).

Figure 3.1a presents the time diagrams showing the operation principle of the azimuth channel RSBN. Figure 3.1b shows the appearance of the terrestrial radio beacons RSBN.

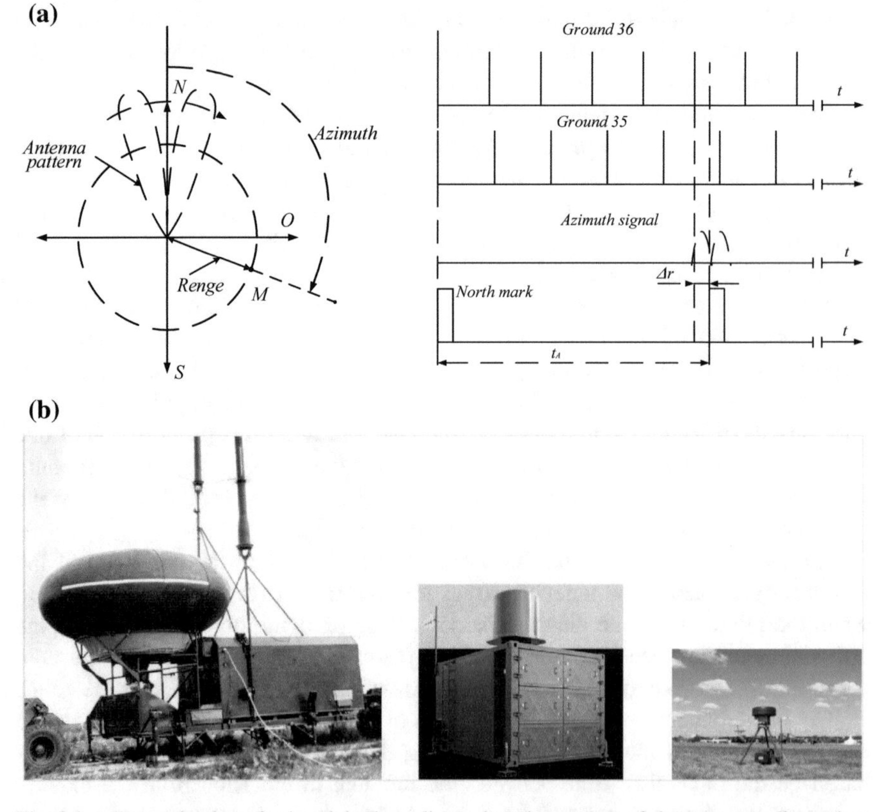

Fig. 3.1 **a** Determination of azimuth in the radio engineering system of short-range radio navigation. **b** External view of the terrestrial lighthouses of the RSBN system

During the development periods, analog methods of processing navigation information were used in the onboard equipment. In recent years, samples of onboard equipment have appeared where the digital navigation methods are used to process the received navigation signal (azimuth and reference channels). Signal processing is performed in specialized calculators that are developed on microprocessors.

In order to increase noise immunity/interference immunity, the antenna system on the moving object has a sector overview of the space. Switching antennas on the moving object are carried out according to the signals of specialized computers.

3.2 TACAN Radio Navigation System

TACAN Radio navigation system (tactical air navigation) has rangefinder and azimuth channels. The rangefinder channel of the system operates in accordance with the request–response principle. Digital delay time meters are used in the receivers of the rangefinders. Taking into account the widely known principle of the operation of such a channel, it is not rational to consider within the framework of this work. The azimuth channel of the system is based on the phase method of operation, which is determined by measuring the phase of the received amplitude-modulated oscillations/waves. A special feature of the azimuth channel is the two-channel method of measurement. At the lighthouse of the TACAN system, a multi-beam directional pattern is formed in the azimuth plane, which is also a cardioid.

Figure 3.2 shows this directivity diagram. It rotates in the azimuth plane at a frequency of 15 rev/s. The indicated shape of the emission patterns and its rotation lead to amplitude modulation of the signal received on the moving object, which is radiated through this antenna. The modulation is performed accordingly at frequencies of 15 and 135 Hz. The signal to rise has a function of the azimuth of the moving object. At a frequency of 15 Hz, this function is single-valued, and the accuracy of determining the azimuth is equal to the accuracy of the phase measurement. At a frequency of 135 Hz, the azimuth is refined (an exact scale).

Reference signals are transmitted with the help of a group of pulses emitted each time when the main or additional maximum of the emission pattern passes through the

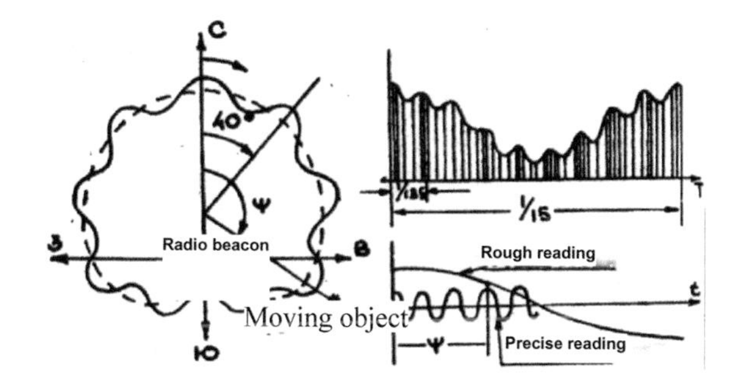

Fig. 3.2 Determination of azimuth in the radio navigation system

Eastern direction of the magnetic meridian. Reference signals are used to synchronize onboard generators with a frequency of 15 and 135 Hz. The reference signals are emitted through the main lighthouse antenna. The doubled root mean square error of azimuth measurement is 1° for the system. Range measurement is provided with accuracy similar to RSBN.

The Micro TACAN system was designed and operated on the basis of TACAN systems for the purpose of determining the coordinates in a group of aircraft related to the leading one.

3.3 Radio Navigation System VOR/DME

The phase method is used to determine the azimuth in the VOR/DME system. At the present time, depending on the structure emitted by the signal lighthouse, the standard Doppler VOR (DVOR) and precision Doppler VOR (PBVOR) are distinguished. In standard VOR, the lighthouse pattern in the horizontal (azimuth) plane has the form of a cardioid. The rotation of this diagram leads to amplitude modulation of the received signal. The phase of the signal to rise of the received amplitude-modulated signal deviates from the phase of the amplitude-modulated signal received in the North direction, by the magnitude

$$\psi_A = \Omega_{вр} t_A \tag{3.2}$$

The azimuth signal in the VOR system is a kind of tape from the signal to rise, the received signal, a sinusoidal voltage of the frequency of 30 Hz (rotation speed of the directional diagram). This signal is called a phase-variable signal. The reference signal emitted by the lighthouse is the frequency-modulated subcarrier of the oscillations/fluctuations, waves/ with an average frequency.

The reference phase signal with a frequency of 30 Hz and a phase independent of the azimuth of the receiving point serves as a modulating voltage. In the North, the phase of the alternating and reference signals of 30 Hz coincide.

The expression for the signal emitted by the lighthouse can be written in the form

$$e = E_m \left[1 + m\sin\left(\Omega_{bp} t\right) - A + m_n \sin(\omega_n t - m_{uм} \cos(\Omega_{bp} t))\right] \sin \omega_н t \tag{3.3}$$

where m is the depth of the signal modulation due to the unevenness/irregularity of the diagram (cardioid);

- Ω_{bp} is the frequency of rotation of the antenna;
- A is the azimuth of the moving object;
- m_n is the depth of modulation of the subcarrier signal;

- ω_n is the signal subcarrier frequency;
- Is the depth of modulation of the frequency-modulated signal;
- Is the frequency of the carrier signal.

Figure 3.3 presents the forms of system signals.

Doppler VOR was designed to reduce the effect of reflected signals on the accuracy characteristics. In the DVOR beacon, the azimuthal signal of the phase variable is transmitted with the help of frequency rather than amplitude modulation. The frequency-modulated signal distorts less due to random variations in amplitude and hence, it has greater interference resistance as compared with the reflections from local objects. In order to eliminate the amplitude distortions, the frequency-modulated signal in front of the frequency detector is fed to the limiter.

The principle of the radio beacon is as follows.

The radio beacon antenna emits continuously unmodulated oscillations/waves through a nondirectional diagram in the azimuth plane. This antenna rotates along a circle of the radius with a circular frequency. In the process of rotation, its distance to the moving object continuously changes with the period of the rotational frequency. The Doppler effect is followed by the oscillation/waving frequency at the input of the receiver on the moving object changes periodically (the received signal will be modulated at a frequency). The phase of this signal depends on the azimuth.

Let us denote R_o the distance from the center of the circle O to the movable object. The distance from the antenna moving along the circumference with the radius to the moving object. Since the relation always occurs, the angle is negligibly small. Therefore, we can write down.

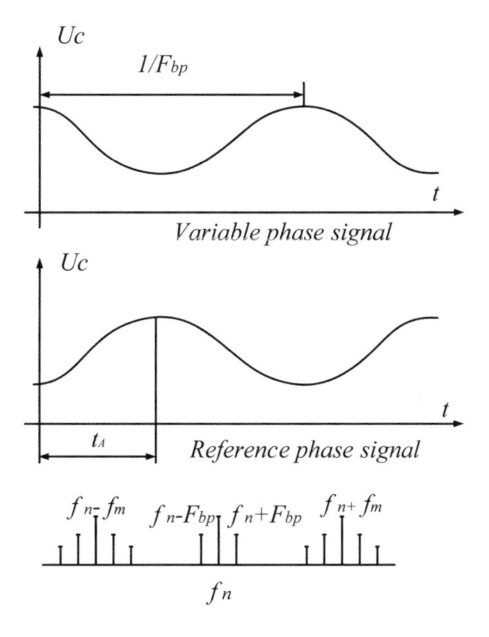

Fig. 3.3 Signals in the system

$$R \approx R_o - r\cos\left(\Omega t - \frac{\pi}{2} - \alpha\right). \tag{3.4}$$

Doppler frequency is being determined by the radial velocity and wavelength.

$$F_A = -\left(\frac{V_R}{\lambda}\right) = -\left(\frac{1}{\lambda}\frac{dR}{dt}\right) = -\left(\frac{r\Omega}{\lambda}\cos(\Omega t - \alpha)\right). \tag{3.5}$$

Therefore, due to the Doppler effect, we can rotate the antenna making it possible to transmit the azimuth signal due to the Doppler effect. Figure 3.4 shows the geometric relationships that determine the work of the DVOR lighthouse. The angle is measured from the direction to the East, so the phase of the reference azimuth signal must be similar to the phase of the frequency modulation in the East direction. When the antenna passes the direction to the moving object, the Doppler frequency is zero, so it causes the need for an additional phase shift of the reference signal while the amplitude of the lighthouse signal at the input of the onboard equipment reaches a maximum.

The radio beacon DVOR was designed in order to receive the azimuth frequency-modulated signal making it possible to use in the onboard equipment a channel for receiving the reference signal of the VOR lighthouse, and, vice versa, for receiving the reference signal DVOR, the channel of the azimuth signal VOR. This means that the kind of modulation of the signals emitted in the DVOR lighthouse is set in places related to the VOR lighthouse. So, the reference azimuth signal represents an amplitude-modulated oscillation/fluctuation, waving by a constant-phase signal with a frequency of 30 Hz in the DVOR lighthouse. In order to obtain the spectrum of a signal with frequency modulation, in accordance with the spectrum that is shown

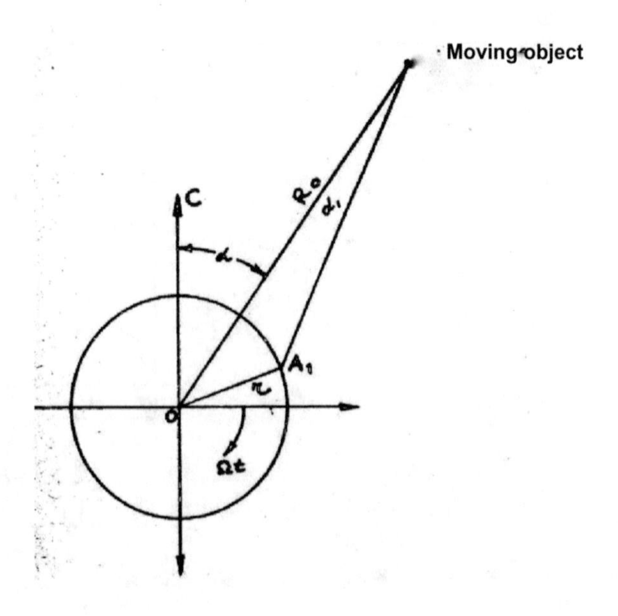

Fig. 3.4 Geometric relations explaining the principle of the lighthouse operation in the system

in Fig. 3.3, one can see that not one but two diametrically located rotating antennas are used. One radiates a signal with a frequency of Hz, and the other signal with a frequency of Hz. Thus, in the air, there is a signal with two lateral frequencies of the spectrum. The carrier frequency of the spectrum is radiated by a nondirectional antenna located in the center of the circle. The radius of the circle along which the antennas rotate is about 6.75 m. Since a 30 Hz antenna is rotated around such a circle. It is difficult, therefore in practice, the commutation of antennas located in a circle is used.

When using a beacon, the azimuthal error decreases approximately 3 times, and in a very difficult/rugged terrain 5–10 times. The doubled average statistical error is 1.5°.

Precision PDVOR beacon gives the advantage to almost completely eliminate the impact of the terrain on the accuracy of the azimuth channel. In the PDVOR, lighthouses information about the azimuth is changed in the same way as in the lighthouse, DVOR, and for the transmission of the reference phase signal, there is a subcarrier of the frequency Hz, frequency-modulated. Advantages can be realized with the use of special onboard equipment. The use of the PDVOR radio beacon and special onboard equipment gives the advantage of getting additional three times of the accuracy of the azimuth intent. The ranging channel of the VOR system operates on the basis of the request–response principle. The equipment is constructed by a digital method of information processing.

Figure 3.5 shows the VOR/DME ground lighthouse.

Fig. 3.5 Appearance of the VOR/DME

3.4 GRAS-Automated System

The GRAS system is a two-channel incoherent phase centimeter-range of the waves intended for high-precision positioning of offshore installations in the production of hydrographic and other tasks. The coordinates of the object are determined by the intersection of two position lines, which are simultaneously measured by the distance to the reference stations. The system is started automatically at intervals of 2, 4, 8, 16, 32, 64, 128, or 256 s specified by the operator. One-time (manual) start is possible when the button is pressed or with the help of signals in the form of pulses of positive polarity with a voltage of 3 … 6 V, coming from the echo sounder/fathometer.

As the unit of distance measurement in case of the phase method, the wavelength of the used oscillation/waving is considered:

$$\lambda = c/f = cT, \tag{3.6}$$

where c is the dissemination velocity of radio waves in a given medium; T and f are the period and frequency of electromagnetic oscillations/waves.

The measured distance can be represented as the sum of the wavelengths that fit in the distance:

$$r = \left(K + \frac{\Delta\varphi}{2\pi}\right)\lambda, \tag{3.7}$$

where K is an integer number of wavelengths. Since the phase meter measures the phase difference within a single period, K is unknown. So $r > \lambda$, it is necessary to resolve the ambiguity of the phase reference.

In RNS GRAS, ambiguity is resolved with the previously described method, i.e., using several measuring (navigation) frequencies:

- F_A—basic frequency, providing single-valued within 100 m $\lambda_A/2$;
- F_B, F_C, F_D—auxiliary the use of which provides with the resolution of ambiguity within 1,000, 10,000, and 100,000 m, respectively.

The nature of the processes that occur in the measuring channels is as follows. The reference/supporting station is installed on the shore at a point with precisely known coordinates, and the ship station emits in the direction of each other electromagnetic waves, which are modulated in amplitude by the corresponding scaling frequencies F_A, F_B, F_C, or F_D. The phase difference between the radiated and received signals on the ship station at the scaling frequency

$$(\Omega_К - \Omega_б)t - \Omega_К\tau \tag{3.8}$$

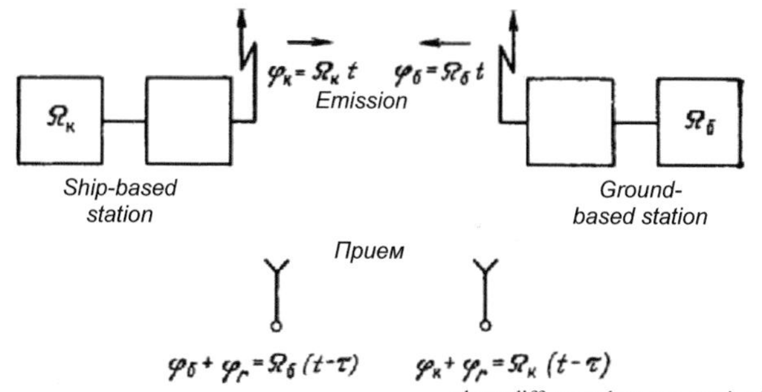

Fig. 3.6 Operations performed at RNS GRAS stations

difference, measured at the coast station,

$$(\Omega_{\text{к}} - \Omega_{\text{б}})t - \Omega_{\text{к}}\tau \tag{3.9}$$

where Ω_k and Ω_b are signal phases at the emission points of the ship and ground stations, respectively.

The operations that are performed at the stations are shown in Fig. 3.6. The difference in the scale frequencies of the transmitters is shown as $\Delta\Omega = \Omega_k - \Omega_b$. These phase differences consist of the common variable of the part $\Delta\Omega_t$ and the different constants $\Omega_{b\tau}$ and $\Omega_{\kappa\tau}$. They differ from each other according to the general law $\Delta\Omega_t$ and have different initial phases proportional to τ.

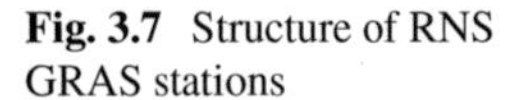

Fig. 3.7 Structure of RNS GRAS stations

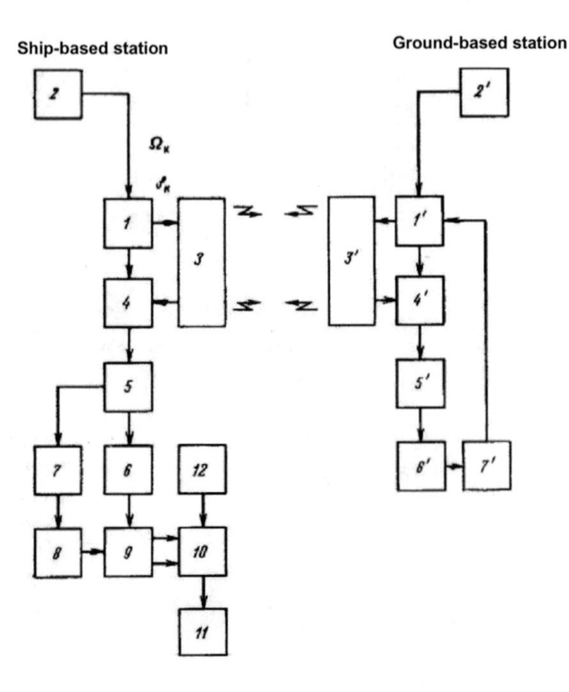

If the phase difference that is measured at the ground station with the help of additional carrier modulation is transferred from the ground station to the ship's station, then considering the delay on the path by τ, the phase of the received signal will be $\Delta\Omega(t-\tau)-\Omega_{\kappa\tau}$. The measured phase is compared with the phase difference measured at the coast station, i.e., $\psi = 2\Omega_{\kappa\tau}$.

The quantity ψ does not depend on the frequency Ω_b. Then, assuming $\Delta\varphi = \psi/2$, the distance

$$r = \frac{\psi c}{2\Omega_{\kappa}} \tag{3.10}$$

Consequently, the distance within half the wavelength is determined in accordance with the measured phase difference ψ at a known velocity and the scale frequency Ω_k. Both the measuring channels of GRAS work do not depend on each other and their functional schemes are almost the same (Fig. 3.7). The nature of the work of each of them is as follows.

At the ship station, the fluctuation of the carrier frequency f_k of the transmitter 1 is phase modulated by fluctuations of the scale frequency Ω_k of the fluctuator 2 and are radiated toward one of the ground stations thus using the antenna device 3. The coast station emits fluctuations with a carrier f_b, which differs from f_k in the value of the intermediate frequency $\Delta f = f_{\kappa} - f_6$. The fluctuations generated by the transmitter 1 are phase modulated by fluctuations of the scale frequency Ω_k of the quartz fluctuate 2, differing by the value of the measuring frequency $\Delta\Omega = \Omega_{\kappa} - \Omega_{6}$. Modulated fluctuations are emitted with the help of an antenna device 3 in the direction of the ship station. Mixer 4 receives FM fluctuations from the ship station and FM natural fluctuations. At the output of the mixer 4′, fluctuations are created at the intermediate frequency with amplitude modulation and a measuring frequency $\Delta\Omega$. The latter is amplified at 5 and detected at 6 (signal).

The same processes can be viewed at the reception section of the ship. At the output 6, a reference signal is being formed. The phase difference between the reference signal and the signal is proportional to the distance between the stations.

In order to transmit the signal from the ground station to the ship station, the signal voltage is frequency-modulated by the subcarrier frequency generator 7, which in addition modulates the carrier f_b of the ground transmitter 1 in phase. This complex fluctuation is received by the ship station. The signal is excerpted using a subcarrier detector 7. The voltage of the subcarrier is detected by the frequency detector 8 and, consequently, the signal is excerpted.

On the selector 9, the voltage of the signal and the reference signal are converted into a pulsed form, which is suitable for further processing, since the intervals between the pulses of the reference voltage and the signal are corresponding to the distance between the stations. The measurements are made on the digital readout device 10, the input of which receives pulses from a special oscillator 12. It passes to the device 11 over the interval between the reference and the signal pulses. The pulse repetition rate of the generator 12 is chosen so that the phase difference is 2π, i.e., one phase cycle is equal to 1000 pulses.

When using the main scale frequency $F_A = 1.5$ MHz ($\Delta\Omega = 2\pi F$), the phase samples expressed by the number of memory pulses will be repeated every 100 m. One pulse is equal to a distance of 0.1 m.

The subsidiary scale frequencies are close to the fundamental and they are chosen so that their differences with the fundamental frequency are 10, 100, and 1000 times less than F, respectively. The resulting sample differences at the frequencies $F_A - F_B$, $F_A - F_C$ and $F_A - F_D$ correspond to the samples at the frequencies $0.1F_A$, $0.01F_A$ and $0.001F_A$.

Thus, each auxiliary/subsidiary scale frequency is used only to determine one significant digit corresponding to the highest discharge of the phasemeter scale. There is a special correction device at the ship station of the system. It performs the logical processing of the measurement of the results, for checking the uniqueness of the account.

Scale frequencies automatic synchronous switching in the system is achieved by sending a signal (code) from the ship station to both coast stations, which starts the switching devices of the ship and shore stations for one switching cycle. To ensure the joint operation of two coastal stations with several (up to 20) ships, emission by each ship station is carried out only during the measuring cycle (0.23 s).

The included ship station operates in a mode in which the transmitter of each measurement channel of the ship station is disconnected from the antenna in between the measurement cycles (determination of the location). At the same time, part of the transmitter power through the directional coupler enters the receiving path thus ensuring the normal operation of the antenna guidance system.

Basic tactical and technical characteristics of GRAS:

– Range (energy) in the presence of sight line	Not less than 60 km
– The minimum distance measured by the system	200 m
– Standard errors in distance measurement	no more than 5 m
– The permissible rate of the measured distances change	up to 50 m/s (100 knots)
– Ship admissible yaw and pitching of the ship in amplitude with a period of not less than 4 … 5 s	not more than ±15°

References

1. Bykov VI, Nikitenko YI (1976) Marine radio navigation devices. M. Transport, 399 pp
2. Shatrakov YG (1990) Radio navigation systems of mobile objects. M. MRP, 92 pp
3. Shatrakov YG et al (1992) Signal processing in radio systems of near navigation. M. Radio and Communication, 370 pp
4. Shatrakov YG et al (1992) Angle-measuring radio engineering landing systems. M. Transport, 362 pp
5. Shatrakov YG et al (1986) Modern systems for short-range radio navigation of aircraft. M. Transport, 402 pp

Chapter 4
Systems of Long-Range Radio Navigation with Ground-Based Stations

4.1 Phase Code Systems of the Long-Wave Range

In order to ensure marine navigation, RNS classified as using long-wave (LW) frequency bands are used: MARS-75, DECCA, LORAN-C, RSDN. Radio waves proportional to this frequency range (kilometer), are well distributed not only over the sea but also over the terrestrial surface. According to the action principle, this is the difference-ranging RNS. In addition, the Decca system is a phase RNS, Mars-75 is an RNS with a multi-frequency signal, and LORAN-C and RSD are pulse-phase RNS. Each of these RNS has its advantages and disadvantages.

The following chapter is devoted to the most commonly used pulsed-phase LW, LORAN-C, and RSDN [1–7].

The systems LORAN-C and RSDN refer to the distance-ranging systems. Their work is based on the pulse and phase methods of measuring the difference in distances. The nature of the impulse method is as follows.

At the receiving point K (Fig. 4.1), the time interval $\Delta t = t_A - t_B$ is measured between the arrival times of two short pulses which are sent by two shore reference stations: Leading A and Driven B.

The distance difference ΔD from the reception point K to the location point of the reference stations A and B is given by the following formula:

$$\Delta \mathrm{D} = D_A - D_B = V_p \Delta t, \tag{4.1}$$

where V_p is the dissemination velocity of radio waves.

The same difference in distance ΔD corresponds to two isolines I-I and I-I′, symmetric with respect to the imaginary O-O-axis (see Fig. 4.1).

For avoiding this uncertainty and the possibility of recognizing the pulses on the screen for some types of transceivers, the driven station B transmits signals with a constant delay

$$t_з = t_в + t_к \tag{4.2}$$

Sauta O.I. et al., *Principles of Radio Navigation for Ground and Ship-Based Aircrafts*, Springer Aerospace Technology, https://doi.org/10.1007/978-981-13-8293-2_4

Fig. 4.1 Nature of the impulse method

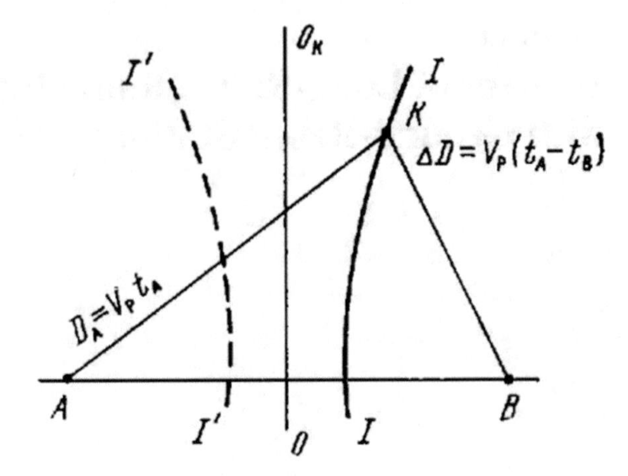

where is the time interval for the base signal traveling through the synchronizing signal; t_k is the code delay.

Therefore, at any point in space, the pulses of the station will arrive earlier than the pulses of the driven. From (4.1) and (4.3) we have

$$\Delta \mathrm{D} = V_p\left(\Delta t - t_з\right) = V_p \Delta t_и \tag{4.3}$$

It is not difficult to show that Δt_u changes from $\Delta t_{u\,\min} = t_k$ from the side of the driven station to $\Delta t_{u\,\max} = 2t_в + t_к$ from the leading one. The difference Δt_i is measured by the impulse method with low accuracy (in a rough manner). More accurately, Δt_u can be measured by the phase method due to the measurement of the phase difference between the oscillations/waves/ that fill the pulses:

$$\Delta\varphi = \frac{2\pi}{T}\Delta t_и \tag{4.4}$$

where T is the period of the oscillations/waves/. It is clear from (4.4) that

$$\Delta t_u = \frac{\Delta\varphi}{2\pi}T. \tag{4.5}$$

Since phase measurements are single-valued only within one period, (4.5) can be written as follows:

$$\Delta t_и = NT + t_Ф + \delta t_Ф \tag{4.6}$$

where N is the integer number of periods of carrier vibrations, unknown as a result of phase measurements; t_ϕ is the measured fraction of the period T; δt_ϕ is the error in measuring the share of the period T.

On the other hand, measuring t_u pulse method, we have

$$\Delta t_{и} = t_0 + \delta t_0 \quad (4.7)$$

where t_0 is a coarse count of the time interval; δt_0 is the measurement error.

From (4.6) and (4.7), we can obtain an unknown number N:

$$N = \frac{\Delta t_0 - \Delta t_\phi}{T} - \frac{\delta t_0 - \delta t_\phi}{T} \quad (4.8)$$

Thus, the clarity condition has the form

$$\delta t_0 - \delta t_\phi < T/2 \quad (4.9)$$

Usually $\delta t_\phi \ll \delta t_0$, therefore the condition (4.9) can be rewritten in the following form:

$$\delta t_0 < T/2, \quad (4.10)$$

that is, to eliminate the multi-value of the phase measurements in the RNS IF, the error in measuring RNP by the impulse method should not exceed half the period of the carrier oscillations/waves/.

The standard LORAN-C IF RNS circuit consists of a master station and up to four driven labeled as W, X, Y, Z, and the emission of which is fixedly synchronized.

All reference stations emit radio pulses of a special shape at a single carrier frequency of 100 kHz. The principle of measuring RNP in the RNS IF is explained in detail in Fig. 4.2.

When phase measurements are performed in the LORAN-C RNS, the zero-point transitions in the positive direction of the high-frequency pulse filling are fixed at the end of their third period, which forms the exact phase count of the RNP Δt_ϕ (Fig. 4.2a).

These measurements use the internal phase structure of impulses. Their instrumental accuracy is of the order of 0.005 … 0.01 of the phase cycle, which is 0.05 … 0.1 mks at the high-frequency filling period $T = 1/f = 10$ mks.

In order to eliminate the multi-value of phase readings, the appearance of an error in the determination of the RNP, equal to nT (n $= \pm 1, \pm 2, \ldots$), is formed from the envelope of the received radio pulses in one way or another (for example, by double differentiating it (Fig. 4.2, lower part) impulse voltage of a special shape.

The zero crossing point of this voltage, thus formed at the leading edge of the received pulse in the region of the third high-frequency filling period, is called the SSP (standard sampling point) and serves to generate coarse samples Δt_0. In this case (4.10), the determination of the RNP will be unambiguous if the error in measuring the RNP by the method δt_0 lies within the range of $\delta t_0 \leq 5$ mks. The methods of forming the SSP are used, without identifying the envelope of the radio signal.

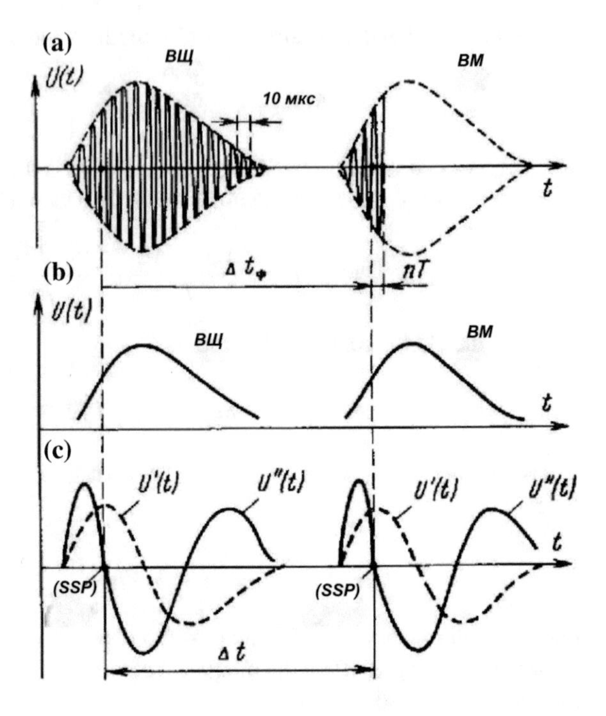

Fig. 4.2 Principle of measurement of RNP in IF RNS

The pulsed nature of the emission of the OS makes it possible not only to ensure the unambiguity of the phase samples, but also gives the advantage to select the surface and spatial signals at the receiving point, thereby making it possible for high-precision phase measurements at considerable distances from the OS.

Obviously, the spatial signals reflected from the ionosphere arrive at the receiving point with some delay δt with respect to the arrival time of the surface signals (Fig. 4.3), which for a pulse duration of the order of 200 mks will result in their interference and the resulting signal with an unstable phase structure, phase measurements will be impossible.

The experimental data of the averaged value of the delay δt of the spatial signals E-1 (single reflected from the E layer) with respect to the surface signals, depending on the time of day and the distance from the OC (Fig. 4.4) allow us to state the following. In the zone of combined action of the surface and spatial signals, making phase measurements within the third period of high-frequency pulse filling, it is possible to obtain a stable phase count of the RNP on the surface signal.

Since spatial signals are significantly absorbed by the lower layer of the ionosphere during the day and are reflected from it only at sufficiently large incidence angles, daytime determinations of RNP up to distances of 600 … 800 miles from the OS are conducted on surface signals. At night, determinations on surface signals are possible at distances up to 600 miles from the OS, since already at these distances the amplitudes of the spatial and surface waves become commensurable. The region is 800–1000 miles in the daytime and 600–900 miles at night a section of the

LORAN-C RNS working area, where the spatial signals E-1 exceed the surface signals in amplitude.

At greater distances, more than 100 miles from the OS, outside the receiving area of the surface signal, it is possible to determine the RNP along the spatial signals E-1, and then correct it with corrections. These corrections for a given pair of stations are applied to the radio navigation charts of the Loran-C system in microseconds and are denoted by one of the letter combinations SS, SG, GS. The first letter refers to the type of signal received from the master station, the second to the slave. Since all stations in the same and nearby circuits of the LORAN-C system emit pulses at a single carrier frequency of 100 kHz, a special organization of their emission is required to ensure spatiotemporal selection of the signals.

Slave stations receive signals from the master station and use them to accurately synchronize the frequency and phase of their own radiated radio pulses. Synchronization is carried out both in the envelope and in the phase of high-frequency filling of radio pulses.

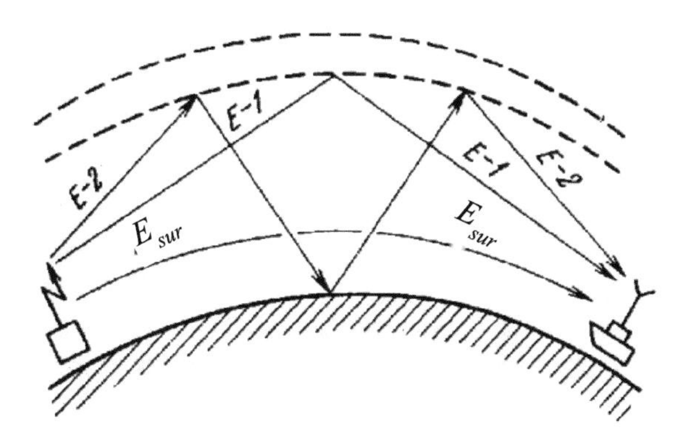

Fig. 4.3 Surface and spatial signals of the RNS RN/IF RNS

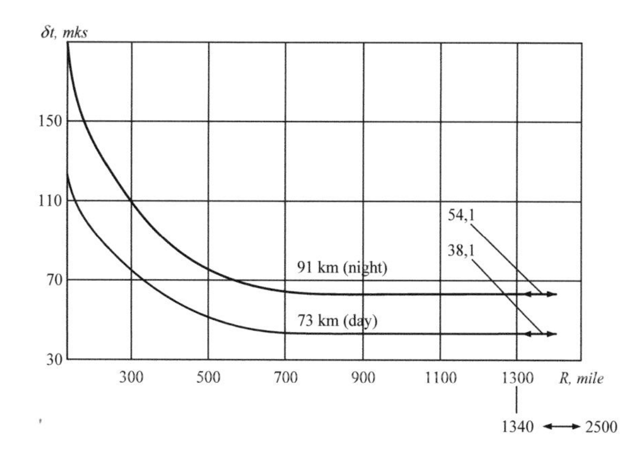

Fig. 4.4 Delay of spatial signals E-1 relative to superficial

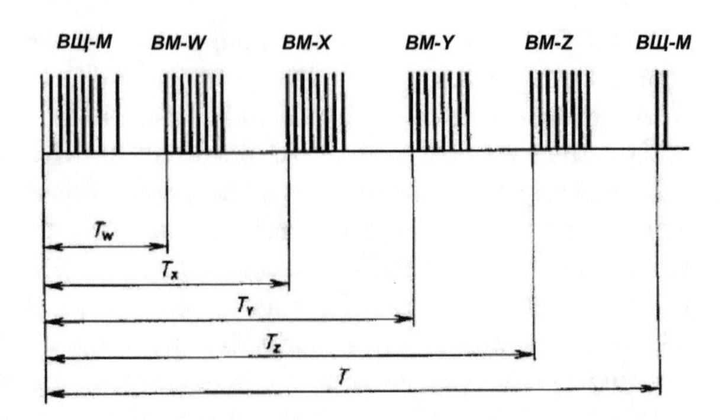

Fig. 4.5 Time diagram of the emission of the RNS LORAN-C

The signals of the slave stations are emitted in a certain sequence, after which the master station again emits, etc. (Fig. 4.5).

Each of the Ns stations emits its signals after a certain protective time interval t_{pi} after the moment of reception of the signals of the station, which in the given order of emissions is the previous one. Then the minimum possible period of emission of signals for the ith circuit is given by

$$T_i \min = \frac{1}{V_p} \sum_{j=1}^{N_c} l_j + N_c t_{\text{pi}}, \tag{4.11}$$

where $\sum_{j=1}^{N_c} l_j$ is the length of a closed line connecting stations in order of priority in the emission of signals; N_c is the number of stations in the chain.

Parameter T_i is used in LORAN-C to identify the signals of a certain chain of stations. Each circuit has its own repetition period T_i, the selection of which is determined when condition (4.11) is satisfied in accordance with the algorithm

$$T_i = (n \cdot 100 - N) \cdot 100 \tag{4.12}$$

where $n = 5, 6, \ldots, 10$—group number; $N = 0, 1, \ldots, 7$—number of the exact value of the repetition period in the group.

The current designation of the RNS LORAN-C circuits is carried out by the so-called frequency parameter, determined through the signal repetition period, as $T_i/10$. In old models of PI the former designation of chains consisting of letter combinations S, SH, SL, SS for group numbers n = 5, 6, 8, 10 is used, respectively, with the addition of number N. For example, the chain 7970 of the Norwegian Sea had the old designation SL3. The conventional designation of a pair of stations consists of the symbol for the chain and the letter designation of a pair of stations. For example, the pair X of the Norwegian chain can be designated as 7970-X or SL3-X. The same notation for the hyperbola of the corresponding pairs is on maps and tables.

The distance of system actions depends on the average power of the signals being received, which is equal to the pulse power of the signal, multiplied by the ratio of the pulse duration to the repetition period.

The increase in the pulse power is limited by the efficiency of the antennas and their electrical strength. The increase in the duration of T does not lead to an increase in the range, since only the initial part of the pulse with a duration of 30 … 40 mks is used for the measurements, which is not affected by the spatial wave. Finally, the increase in the average power due to the reduction of T is limited by the condition (4.11).

That is why for medium-size and long-term development, consequently, the application of the following action is accepted in the RNS LORAN-C. The optical station is not a single impulse for the T period, and the whole series is composed of eight impulses at the stations and the number of points (see Fig. 4.5).

The duration of the pulse in the package is 135 mks, the interval is 1000 mks in the package with the pulse in the 9th marker at the point of departure from the last impulse 8 to the 2000 mks. When emitting eight pulses, the total signal power is increased by 4 times compared with the power of a single pulse.

The initial phase of the high-rise impulses in the pads can be changed to 180° in correspondence with the phase encoding, given in Table 4.1.

The law of phase encoding of packets of all slave stations is the same and differs from the laws of encoding the impulses of the master station, which ensures the search and recognition of its signals in automatic receiver. The period of Tk encoding includes two periods of T of packet sequence: odd and even.

The phase coding makes it possible to substantially eliminate in onboard receiver the effect of repeatedly reflected from the ionosphere previous impulses in the package for subsequent surface, and also expands the spectral components of the signal within the operating frequency band of the system, which is an additional means of combating noise interference.

The accuracy of maintaining the radio navigational field in the working area of the IF RNS is determined to a large extent by the stability of the fluctuations of supporting generators of OS, which is determined by the cesium standards of frequency.

The emission of all LORAN-C circuits is synchronized with the World Coordinated Time (UTC). Each circuit consists of one or two control points, which periodically checks the accuracy of mutual time shifts in the emission of signals, and generates control signals for reducing the timescales of the OS.

Table 4.1 The initial phase

Emission	Leading station	Driven station
Impulses	1 2 3 4 5 6 7 8	1 2 3 4 5 6 7 8
Even series	+ + − − + − + −	+ + + + + − − +
Odd series	+ − − + + + + +	+ − + − + + − −

Note The "+" symbol conditionally corresponds to the phase bearing fluctuations taken as 0°, and the "–" sign to the change of this phase by 180°

If the synchronization of any station exceeds the established limits (0.1 mks) or the violation of the form of emitted impulses, the emergency station transmits warning signals to the consumers at the command of the control unit. This is either amplitude manipulation by the first two impulses in the packet of the emergency slave, or by manipulation of the 9th impulse of the lead packet in accordance with the Morse code, which determines the emergency station. Cycle manipulation is 12 s.

With the length of the baselines of 600 … 1200 miles, the pulsed radiated power of the transmitters of the stations, depending on the nature of the underlying surface in the working area, is 165 … 2100 kW. Antenna OS systems are umbrella antennas-masts up to 412 m in height or antenna-cloths in the form of a square with a side of 425 m.

The radio navigation charts of the RNS LORAN-C are maps in the Mercator projection with the pairs of stations of hyperbolas applied on them, the signals of which are received in the area covered by the map.

Hyperbolas are digitized in microseconds and are usually performed after 50, 100 μs, depending on the scale of the map. The observed place is found as a point of intersection of the corresponding hyperbolas. If necessary, the required hyperbolas are found by interpolation using a ruler with a uniform scale. If the signals of different pairs of IF PNS stations are received at different times, the first line of the position should be taken to the place where the second one is taken, moving it along the course by the amount of travel during the time between observations.

When receiving signals on a spatial wave, the value of the hyperbola taken with the receiver should be corrected by the correction for the difference in the course of the surface and spatial radio waves. Daytime corrections are indicated by the letter D, night—by the letter N. The daily corrections are always less than the night ones. As mentioned above, there are probably three types of corrections: SS, GS, and SG.

Thus, near the point with coordinates $\varphi = 32°$ N and $\lambda = 72°$ W, for example, located on the fragment of the map N21008-LC (Fig. 4.6), to the samples of the RNP (hyperbola) from the pair 9960-X when receiving signals of the SS type in the daytime a correction of 3 mks should be given, to the RNP samples from the pair 9960-X when receiving signals of the GS-type during the day-correction +41 mks.

If the vessel is at the point with coordinates by calculation $\varphi_c = 31° 50.4'$ N; $\lambda_c = 70° 59.0'$ W, and the corresponding measured RNP values from a pair of stations 9960-XTSs = 26,089.0 mks, from a pair of stations 9960-YTsg = 40,110.4 mks, then the corrected RNP values will be, respectively, T = 26,086.0 mks and T = 40,151.4 mks. After that, the construction can be used to obtain the vessel's observable coordinates (see Fig. 4.6).

We will consider the method of obtaining the vessel's place by signals of the RNS LORAN-C using the tables. A fragment of such tables for two pairs of stations X, Y of chain 9960 (SS4) is given in Tables 4.2, 4.3.

Arguments for entering the tables of this pair of stations are the value of the corrected RNP value defining the hyperbola and two close to the calculable and satisfying the condition $\varphi_1 < \varphi_c < \varphi_2$ of latitude. The value of Δ in the tables is the change in longitude at the point of intersection in hundredths of a minute by 1 mks.

For the considered example, taking into account Tables 4.2, 4.3 we have:

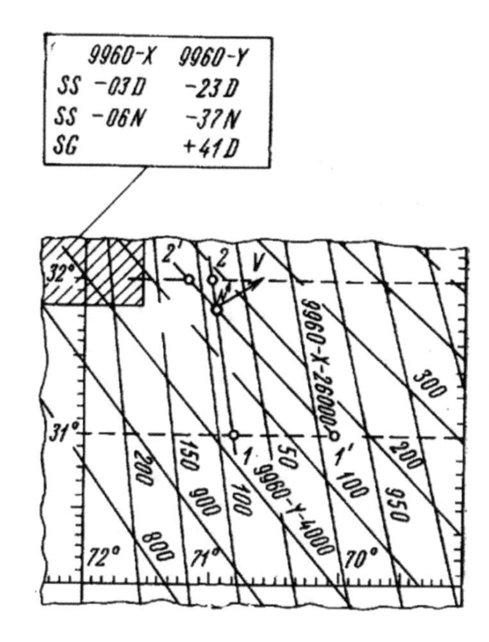

Fig. 4.6 Accounting for amendments to the IF RNS LORAN-C

Table 4.2 Couple 9960-X (SS4-X)

T	– – – – –	26,080		26,090		– – – – – –
φ° N		λ	Δ	λ	Δ	
31		70° 46.5′	45	70° 51.1′	45	
32		70° 57.2′	41	71° 01.4′	41	

Table 4.3 Couple 9960-Y (SS4-Y)

T	– – – – –	40,150		40,160		– – – – – –
φ° N		λ	Δ	λ	Δ	
31		70° 04.1′	−37	70° 00.3′	−37	
32		70° 11.2′	−33	71° 07.8′	−33	

Table 4.4 LP-9960-X, Tx = 26,086; T = 26,080

No point	φ_T	λ_{table}	Δ	$T_x - T$	$(T_x - T)$	λ_T
1	31° 11′	70° 46.5′	+0.45′	+6	+2.7′	70° 49.2′
2	32° 11′	70° 57.2′	+0.41′	+6	+2.5′	70° 59.7′

Applying the coordinates of the points obtained on the waypoint map and passing through the points 1 and 2 the line of positions I of the LP, through the points 1′ and 2′-II LP, we get the observable coordinates of the vessel φ_0, λ_0 (Tables 4.4, 4.5).

Tables are for the case of reception of signals on surface waves. The spatial waves are accounted for by corrections, also chosen from a table, usually placed before the

Table 4.5 LP-9960-Y; TY = 40,151.4; T = 40,150 (Table 4.5)

No point	φ_T	λ_{table}	Δ	$T_y - T$	$(T_y - T)$	λ_T
1	31°	70° 04.1′	−0.37	+1.4	−0.5	70° 03.6′
2	32°	70° 11.2′	−0.31	+1.4	−0.4	71° 10.8′

coordinates tables of hyperbola points. Arguments for entering the correction table are the computable coordinates of the location of the moving object.

It is clear from the foregoing that the location of a moving object with the help of radio navigation charts (tablets) is faster than with tables. At the same time, tables allow you to find a place immediately on the map.

Working with automatic PI requires, as a rule, the introduction of corrections that take into account the deviation from the propagation speed of radio waves from the true one, and a number of other factors.

To determine the location by IF RNS LORAN-C, two hyperbolas are usually used. Consequently, the radial mean square error in determining the location depends on the errors in determining the RNP and the geometric factor.

The radio navigation parameter for the long-wave pulse-phase PHS is the time interval between the moments of arrival of signals from the master and slave stations.

Investigations established that when receiving signals on surface waves, the total root mean square error $\delta_{\Delta t} = 0.4 \ldots 0.5$ mks at distances up to 800 … 1000 miles, when receiving signals on spatial waves; $\delta_{\Delta t} = 2 \ldots 3$ mks at distances over 1500 … 1800 miles from the stations.

When measuring the time interval only by the impulse method, that is, on the envelope of the radio impulse at surface waves, we obtain $\delta_{\Delta t} = 2 \ldots 3$ mks, on spatial waves $\delta_{\Delta t} = 4 \ldots 5$ mks and more.

The results of calculating the radial mean square error of determining the location M, for $\eta = 1.64$, are as follows:

on surface waves M = 196 … 246 m;
on spatial waves M = 984 … 1476 m (in the absence of systematic errors due to incorrect determination of the type of correction for spatial waves).

Practice shows that usually, the errors in locating with the help of the Loran-C RNS with a probability of 0.95 are 0.2 … 0.3 miles in the daytime on surface waves under favorable conditions and 0.5 … 1.0 miles at night at extreme distances.

Systematic errors, determined by ignorance of the true conditions of propagation of radio waves, have 2 … 3 μs, which is an order of magnitude greater than the random errors and has the greatest influence on the accuracy of locating.

These systems are widely used in aviation.

4.2 Super-Long-Wave Range Systems

The work on the creation of a global RNS on SLW began abroad after the Second World War and by the mid-1950s the projects "Delpak" (1954, Great Britain) and "Redaks Omega" (1957, USA) were developed. There was an improved project combining the positive qualities of the two abovementioned works and was named Omega.

Beginning from 1966, the Omega RNS entered the number of operating radio navigation systems. In 1980, eight stations of this system were installed. They provide a definition of the location on the entire surface of the Earth.

In the Russian Federation, a similar domestic airborne navigation system was developed and operated by SLW RNS Route.

The Omega system is a phase difference ranging (hyperbolic) radio navigation system designed for navigational determinations on the ground, in the air, underwater, and under ice at any time of the day and year, in all weather conditions and, what is extremely important, at any point in the World Ocean. Therefore, it was called the global system.

The RNS Omega, which has international status, includes eight coastal reference stations (OS): in Norway, the United States, the Hawaiian Islands, Liberia, Japan, Reunion, in the south of Argentina, and in Australia.

The designations of the OS, their coordinates in the International Geodetic System WGS-72, the type of antennas used, as well as those responsible for normal operation are presented in Table 4.6.

The basic distance between the stations amounts to about 5,500 miles on average. To determine the location of the vessel in any part of the world requires only six stations. Two more stations are required to obtain redundant information and the possibility of decommissioning the stations for repair. Anywhere in the world, you can receive signals from four stations and receive six position lines with angles of intersection of at least 60°.

Very low frequencies 10 … 14 kHz ($\lambda = 22 \ldots 0.30$ km) of radio waves are used in the RNS Omega, so the system is called super- long-wave range. Radio waves of this range have favorable conditions for their stable propagation over long distances (up to 10,000 km), they are little absorbed in the lower layers of the atmosphere and penetrate well into relatively large depths (10 … 30 m) underwater.

The reference stations emit phase-locked signals with a temporal (for operating frequencies) and frequency (for individual frequencies) channel separation in accordance with the emission pattern (Fig. 4.7).

The main operating frequency is $f_0 = 10.2$ kHz. Additional operating frequencies $f_1 = 11.05$ kHz, $f_2 = 11.1/3$ kHz, $f_3 = 13.6$ kHz are emitted to resolve the multi-valued phase samples. For this purpose, on PI board, measurements can be made at the difference frequencies:

$$\Delta f_{30} = f_3 - f_0 = 3.4\,\text{kHz};$$
$$\Delta f_{20} = f_2 - f_0 = 11331/3\,\text{Hz};$$

Table 4.6 Reference stations

Station	Notation	Coordinates of the OS of the RNS Omega	Antenna	A responsibility
Norway (Aldra)	A	66° 25′ 12.62″ 13° 08′ 12.52″	Valley	Norwegian management
Liberia (Painsworth)	B	6° 18′ 19.11″ 10° 39′ 52.40″	Grounded mast 420 m	USA under the contract[a]
Hawaii (Cameron)	C	21° 24′ 16.78″ 157° 49′ 51.51″	Valle	Coast security USA
Sev. Dakota (LaMoore)	D	46° 21′ 57.29″ 98° 20′ 08.77″	Isolated mast 360 m	Coast security USA
Reunion (Mafate)	E	20° 58′ 27.03″ 55° 17′ 23.07″	Grounded mast 420 m	France, Navy
Argentina (Treplev)	I	43° 03′ 12.89″ 65° 11′ 27.36″	Isolated mast 360 m	Argentina, Navy
Australia (Woodside)	G	38° 28′ 52.53″ 146° 56′ 06.51″	Grounded mast 420 m	Australia, Department of Transportation
Japan (Tsushima)	H	34° 36′ 52.93″ 129° 27′ 12.57″	Isolated mast 450 m	Japanese Shipping Safety Organization

[a]The US contract until 1982, then the responsibility is taken by the Ministry of Industry and Transport of Liberia

$$\Delta f_{21} = f_2 - f_1 = 2831/3\,\text{Hz};$$

A precise grid of hyperbolas is created on the frequency of $f_0 = 10.2$ kHz, which corresponds to $\lambda = 29.4$ km. That is why, the vertical tracks on the base are $d_0 = \frac{\lambda_0}{2} = \frac{29.4}{2} = 14.7\,\text{km} = 8\,\text{miles}$

Consequently, for the bases used in the RNS Omega, about 700 tracks are obtained, that is, the system has a large polysemy.

To resolve polysemy, it is necessary to determine by some other methods the position of the vessel with an error of not more than half the width of the track, i.e., up to 4 miles. In some cases, this accuracy can be obtained by counting. However, in the ocean, this cannot always be done, so the method of binding, based on the numbering, for resolving the polysemy of the Omega RNS is of little use.

Measurements at the difference frequencies make it possible to increase the single-valued phase counting zone to the corresponding values of the rough tracks at the difference frequency:

$$\Delta f_{30} \ldots d_{30} = 3d_0 = 44.1\,\text{km} \approx 24\,\text{miles};$$
$$\Delta f_{20} \ldots d_{20} = 9d_0 = 132\,\text{km} \approx 72\,\text{miles};$$
$$\Delta f_{21} \ldots \Delta f_{21} = 36d_0 = 4d_{20} = 529\,\text{km} \approx 288\,\text{miles};$$

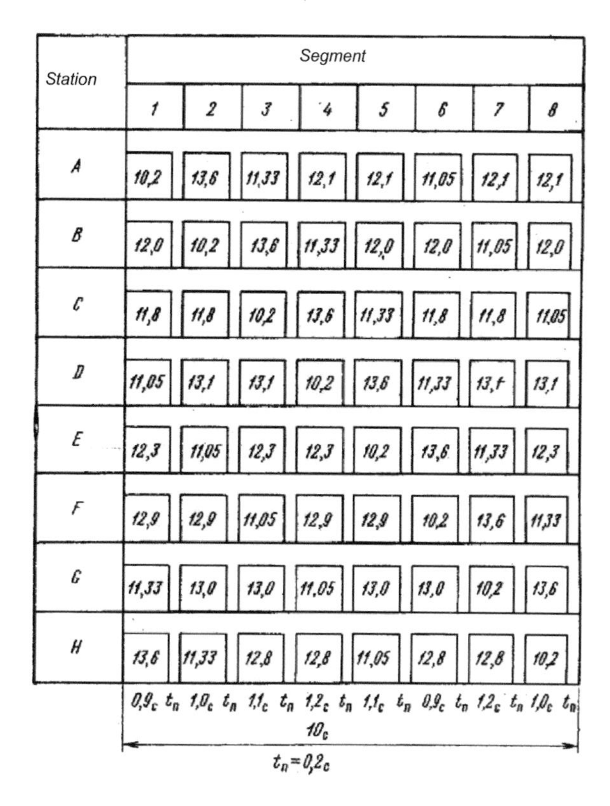

Fig. 4.7 Diagram of emission of supporting stations of the RNS Omega

To resolve the polysemy, as a rule, the frequencies f_0, f_2, f_3 are used. The frequency $f_1 = 11.05$ kHz is included in the emission diagram for the resolution of polysemy of phase readings in the case of navigating objects during search and rescue operations and for air navigation in order to expand the zone of unambiguous phase count and reduce the accuracy requirements for autonomous navigation aids that provide counting.

The duration of a pause between radio signals is $t_n = 0.2$ s, which eliminates the possibility of overlapping signals from different stations at any point in the work area, for example, due to differences in signal delays in the propagation of radio waves on the routes from the respective stations. In addition, the adopted $t_n = 0.2$ s reduces the requirements for the accuracy of the receiver switch configuration—the permissible error of the installation is of the order of 0.1 s. The emission cycle of the OS is 10 s. In this cycle, the radio signals of the operating frequency always occupy four segments out of eight (see Fig. 4.7). In the other four segments, each station emits individual frequencies, which serve to overtone the stations and transmit additional information. For example, the possibility of using individual frequencies for transmission of control signals within the system and transmission of a special time code that allows unambiguous binding of the user's local hours was discussed. In some aircraft receiver s, the signals of the Omega RNS on individual frequencies are used to determine the location together with the signals of the VLF stations of the United States Navy.

The timescale (epoch) of the RNS Omega is noted at each station at the moments of simultaneous passing of the oscillations of all four produced navigation frequencies through zero in the positive direction, which are repeated with the frequency of the general subharmonic of the data $F_{cr} = 850/3$ Hz (period $T_{cz} = 1/F_{cz} = 60/17$ ms is a multiple of the interval of 30 s). The accuracy of generating navigation frequencies at each station is provided by an atomic frequency standard consisting of four cesium generators with relative instability $\Delta f/f = 10^{-12}$.

The epoch of the Omega RNS is synchronized with the World Coordinated Time (UTC), controlled by the atomic timescale of the US Navy Observatory (USNO). It should be noted that the TC scale is periodically shifted to the UT-2 scale, which takes into account the rotation of the Earth around its axis. All of the time shifts in the UTC scale of the Omega RNS are conducted since the beginning of its operation annually from December 31 to January 1. The Omega RNS does not take into account these jumping seconds, so in 1986, it was ahead of the UTC World Coordinated Time for 15 s, and this difference is growing with every annual jumping second.

There is no division of stations into master and slave stations in the RNS Omega, which is a practical means of achieving redundancy in navigational measurements, since any two stations of the system can form a navigation pair and create a hyperbolic isoline. Thus, the 8 stations of the system allow forming a total of 28 pairs of stations.

Independence of station operation is achieved by using the abovementioned atomic frequency standards at each of them, synchronized both with each other (internal synchronization) and in respect to the World Coordinated ITS time scale (external synchronization), taking into account the definition and features of the Omega RNS epoch above.

To ensure the synchronization of signal from coastal stations in close proximity to them, a network of control points (CP) with an accurate geodetic reference is deployed. At each CP, the phase differences of the signals of the own station are measured on three navigation frequencies relative to the signals of the remaining stations during the ideal cycle. The CP also measures the discrepancies of the station's own timescale and the TC scale.

All this information is transmitted through the communication channels to the System Management Center (SMC). In the SMC, a special program on high-speed computers is used to calculate the approximated values of the discrepancies of the scales of each OS relative to the RNS average timescale and an estimate of the position of the average timescale relative to the World Coordinated Time scale. On the grounds of the obtained results, the position control signals of the OS scales are generated. The accuracy of internal and external synchronization systems is 1 mks (rms value).

The emission power of the reference stations of the Omega RNS is at least 10 kW at a frequency of 10.2 kHz. The transmitting antennas of the reference stations have rather large and complex designs that largely determine the cost of the reference stations. These are either umbrella antennas (with an isolated or grounded mast about 420 m high), or antennas of the valley type, which are wires stretched across the gorge (valleys).

The initial information for obtaining the craft's position in the Omega RNS are the measured receiver values of the phase difference of radio signals of the corresponding frequency for the selected pairs of reference stations—hyperbolic coordinates. At that, as a consequence, methods of obtaining the location, traditional for distance-ranging radio navigation systems, such as LORAN-C, RS-10, etc.

This is, first, a graphical-analytical method, when in the case of nonautomatic receiver s, radio navigation charts with hyperbola grids and tables with coordinates of hyperbola points for a given pair of stations are used for subsequent manual processing.

At that, second, an analytical method, when automatic algorithms are used to convert hyperbolic coordinates into geographical or rectangular coordinates with the use of mathematical methods with automatic calculation and correction.

Getting the location of the craft using a radio navigation map (pad) is simple enough. To do this, we need to find the intersection point of the hyperbolas of the pairs of stations for which the phase differences are measured.

To determine the location for the RNS Omega when navigating in the seas and oceans, radio navigation maps (pads) are used. The scale of the pads is 1: 2,000,000. A fragment of one of these pads is shown in Fig. 4.8.

Hyperbolas on the pads have different colors and correspond to the boundaries of the tracks at a frequency of 10.2 kHz. Small-scale pads usually digitize tracks, the numbers of which are multiples of three. At the same time, these tracks are the boundaries of the zones obtained at a frequency of 3.4 kHz; tracks, multiples of nine—the boundaries of wider zones corresponding to a frequency of 1.13 kHz.

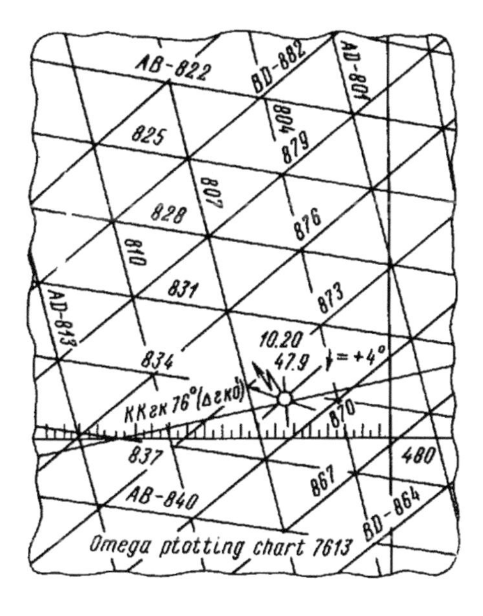

Fig. 4.8 Detail of the radio navigation map (pad) of the RNS Omega

Hyperbolas are assigned letter indices of the corresponding pairs of stations. For example, on the chart above (see Fig. 4.8), the hyperbolas have the indices of pairs AB, AD, and BD.

Tracks have three-digit identification numbers, which are obtained using the receiver track counter. The line of position on the track is found by interpolation over the fractional part of the phase difference, expressed in hundredths of the phase cycle. Hyperbole, which coincides with the normal to the database, is assigned 900. The track numbers on the counter will change with the movement of the craft. In this case, moving along the increasing track numbers means furthering the distance from the station indicated by the first letter and approaching the one indicated by the second letter, and vice versa.

Maps and tables for determining the location for the RNS Omega are calculated for the average daily height of the ionosphere in case of radio waves propagation over land Vp = 300,574 km/s. With this choice of Vp, the propagation of radio waves along the illuminated path, the deviation $\Delta V_D = V_{\Phi\text{E}} - V_p$ of the actual phase velocity VΦD from the calculated value Vp will be close to zero, while the propagation along the shadow path—the maximum: $\Delta V_H = V_{\Phi H} - V_p = \Delta V\text{max}$.

Graphically the change in the value ΔV during the day can be represented in the form of a trapezoid (Fig. 4.9).

The decrease in the phase velocity at night leads to an increase in the number of phase cycles that fit into this propagation track. Consequently, moving from day to night, the phase count at a given point will increase. To compensate for this, entering a correction for the negative character is needed.

Theoretical and experimental studies have shown that changes in phase velocity in relation to the season and time of day have good repeatability, which makes it possible to calculate them and compensate for them in determining navigation. Amendments depending on the radio waves propagation conditions are issued for 26 regions of the globe (Fig. 4.10). Each of the areas is restricted to 30 and 45° latitude and 60° longitude. In the polar regions $\Delta\varphi = 15°$, $\Delta\lambda = 180°$. The amendments are issued in the form of tables in the Omega propagation correction tables for each station and each area for different frequencies.

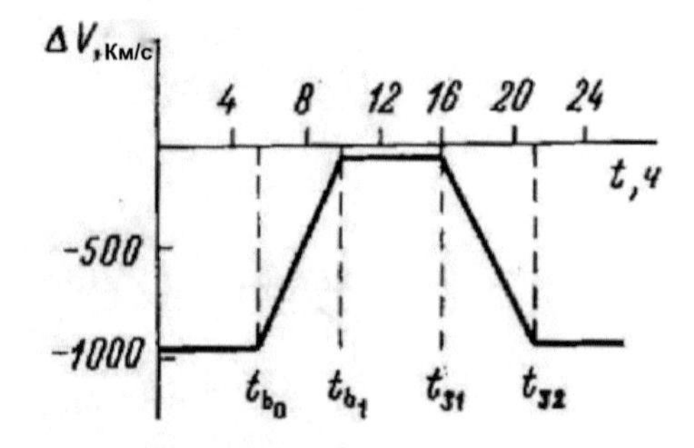

Fig. 4.9 The graph of the value change ΔV

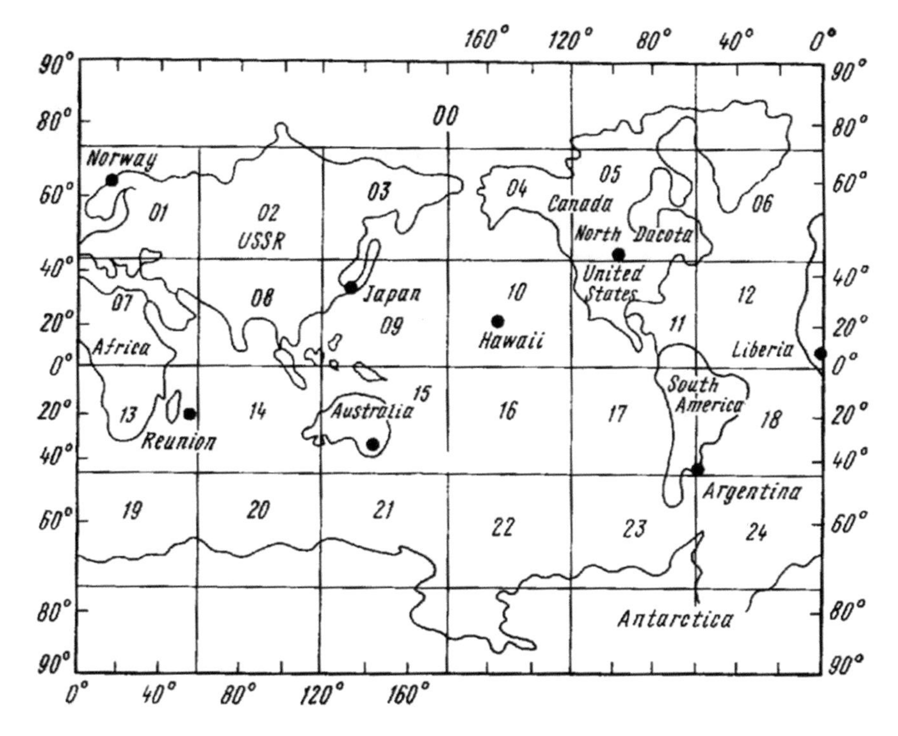

Fig. 4.10 Numbers and boundaries of amendments areas for the propagation of radio waves for the RNS Omega

The arguments for entering in the tables are

the craft's calculated coordinates;
date with half a month accuracy;
Greenwich Mean Time every hour.

Corrections for a pair of stations are determined as the difference in the corrections of the corresponding stations, for example, $\Delta_{A-B} = \Delta_A - \Delta_B$.

Correction tables are being corrected. Extracts from the tables of corrections of stations A, B, and D for $f_0 = 10.2$ kHz are given in Table 4.6. The values of the corrections in it are given in hundredths of the phase cycle (centi-cycles).

Along with the pads for determining the craft's location, the Omega RNS tables containing the coordinates of the hyperbola points of certain pairs of stations are also widely used to determine the location.

Tables are compiled for the indicated 26 regions of the globe (see Fig. 4.10). Four–six tables are compiled for each of the regions. One table is suitable for only one pair of stations.

Tables are made for hyperbolas formed when receiving signals at $f_0 = 10.2$ kHz. There are also specific tables for other operating frequencies.

Let us illustrate the use of the tables of the Omega RCS with the example, using the fragments in Tables 4.8, 4.9 of this chapter. These tables are entered with the corrected phase difference values and latitudes (or longitudes) close to (φ_c, λ_c), and then the longitudes (or latitudes) of the two hyperbola points are chosen. These points are transferred to the enroute map and a line of position is drawn over them (Fig. 4.11).

The value Δ in Tables 4.7, 4.8, 4.9 represents the change of the desired coordinate by one phase cycle, the letter T in the Omega table denotes the phase difference (Table 4.10).

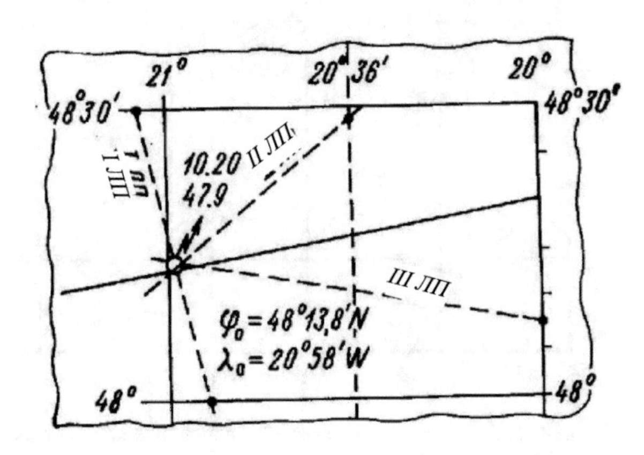

Fig. 4.11 Determining the location by means of the Omega RNS tables

Table 4.7 Location 48° N 18° W

Station A Norway	
	00 01 … 08 09 10 11 12 … 23 24
1–15	−35 −37 … −26 −11 −6 −18 −18 … −36 −35
Station B Liberia	
1–15	−64 −65 … −11 −6 −8 −7 −5 … −65 −61
Station D North Dacota	
1–15	−61 −63 … −64 −58 −47 −39 −32 … −59 −61

Table 4.8 APEA 06 PAIR B–D

B–D 871			B–D 872			T
	′	Δ′		′	Δ′	Long
48	10.4	11.0	48	21.4	110	21°
48	41.4	10.9	48	52.3	100	20°

Table 4.9 APEA 06 PAIR A–B

A–B834			A–B835			T
	′	Δ′		′	Δ′	Long
48	12.4	−8.7	48	03.7	−8.6	20°
48	18.4	−8.7	48	09.6	−8.7	21°

Table 4.10 Latitude and longitude definition

No. points	φ_m	$\lambda_{табл}$	Δ	$T_x - T$	$(T_x - T)$	λ_m
II LP B–D T = 871.25; *T* = 871						
1	20°	48° 41.4′	+10.9′	+0.25	+2.7′	48° 44.1′
2	21°	48° 10.4′	+11.0′	+0.25	+2.7′	48° 13.1′
III LP A–BT = 834.48; *T* = 834						
1	20°	48° 12.4′	−8.7′	+0.48	−4.2′	48° 08.2′
2	21°	48° 18.4′	−8.7′	+0.48	−4.2′	48° 08.2′

Before analyzing the accuracy characteristics of navigational determiners and the factors affecting them, it is obviously advisable to consider how the working areas are defined in the Omega RNS. This is closely related to the peculiarities of the propagation of extra long waves, and in particular, to the concept of the modality of their propagation and its influence on phase measurements. Taking into account the properties of the polymodal propagation, four areas are distinguished according to the distance from the coast stations:

I area—up to 30 miles from the emitter. In this case, the consumer of navigation information is within the so-called induction zone. The electromagnetic field of the emission has not yet been formed, this zone is not used for navigation definitions.

II area—up to 220 miles from coast stations. The waveguide nature of the oscillations has not yet been established, the electromagnetic field is being formed only by a surface signal. This area, under certain restrictions, can be used for navigational definitions.

III area—up to 1100 miles from stations. The area of interference of different modes' oscillations. Phase measurements are unreliable, can lead to significant errors in determining the location. Hyperbolic isolines on maps of the Omega system are drawn with dashed lines, which indicate the limited use of this area for the purposes of radio navigation.

IV area—up to 6000 miles, the actual working area of the RNS Omega. The use of receiver in this area requires taking into account corrections for the daily and seasonal course of the phase caused by changes in the parameters of the natural waveguide.

To quantify the accuracy of the location of the RNS Omega we will use a circular error. The error in determining the location of a vessel depends on the error in determining the difference in distances and the value of the geometric factor.

The error in determining the distance difference $\sigma_{\Delta D}$ in its turn is determined by the error in measuring the phase difference $\sigma_{\Delta\psi}$;

$$\sigma_{\Delta D} = \lambda_0 \sigma_{\Delta\psi} \tag{4.13}$$

where λ_0 is the wavelength of the navigation frequency f_0, on which phase measurements are made, and $\sigma_{\Delta\psi}$ is the error in measuring the phase difference in fractions of the base cycle.

The error in measuring the phase difference is determined by the following components:

error of synchronization of stations (does not exceed ±0.01 ph.c. per day);
instrumental error of receiver (about ±0.02 ph.c.);
error due to the inconsistency of propagation conditions (at daytime ±0.03 ph.c., at nighttime—about ±0.05 ph.c.).

Having adopted the hypothesis of randomness and independence of the components of the error, we have:

at daytime $\sigma_{\Delta\psi} = \pm 0.04$ ph.c., $\sigma_{\Delta D} = 29.4\,\text{km} \cdot 0.04 = \pm 1.17\,\text{km} = \pm 0.63$ miles;

at nighttime $\sigma_{\Delta\psi} = \pm 0.06$ ph.c., $\sigma_{\Delta D} = 29.4\,\text{km} \cdot 0.06 = \pm 1.77\,\text{km} = \pm 0.96$ miles;

The errors in determining the position of the craft along two lines of position with the minimum possible value of the geometric factor $\eta = 0.7$ are

at daytime $M = 0.7 \cdot 1.17 = 0.82\,\text{km} = 0.44$ miles;

at nighttime $M = 0.7 \cdot 1.77 = 1.23\,\text{km} = 0.66$ miles;

The minimum possible value of η will be at the point of intersection of two orthogonal baselines. At this point, the viewing angles of the bases $\gamma_1 = \gamma_2 = 180°$ and $\theta = 90°$, therefore,

$$\eta = \frac{1}{2 \cdot 1}\sqrt{1+1} = 0.7 \tag{4.14}$$

In the transition period (between day and night), the error in measuring the parameter $\sigma_{\Delta\psi} = \pm(0.07\text{–}0.08)$ ph.c., so the accuracy of determining the location in this period is decreasing.

As mentioned before, Omega is essentially the first system of radio navigation, the development of which specifically provided for the use of redundancy of navigation information, redundancy of position lines to improve the accuracy and reliability of position determinations.

In addition, it is interesting to note that in determining the position of the craft with three lines of position, the position error is approximately 25% less than the error obtained from the two lines of position.

The above estimates of the accuracy of location suggested that the observations are not correlated. However, this is not always true. If the same station is used to measure the phase difference with the other two, a correlation is possible between the obtained position lines.

Table 4.11 Cross-correlation coefficient

Number of pairs counted	Pairs of stations	Distance miles	ρ	Note
362	$X = A - B$	$D_A = 975$	$\rho_{xy} = +0.05$	A close, with the same character
	$Y = A - D$	$D_B = 2765$	$\rho_{xz} = -0.25$	B with different characters
	$Z = B - D$	$D_D = 3600$	$\rho_{yz} = +0.88$	D far away, with the same character

The value of the correlation coefficient p for the Omega RNS varies in a wide range ($0 < p < |1|$). Thus, in Table 4.11, experimental data on the determination of the cross-correlation coefficient are given.

Numerous experimental evaluations of the accuracy characteristics of the Omega RNS conducted in the Russian Federation and abroad show that the real radial mean square errors in determining the location are 3 … 4 miles, depending on the geometric factor. There were cases of larger errors that were observed during the period of sharp changes in the corrections.

A significant decrease in the accuracy of the location of the Omega RNS may occur during the onset of ionospheric magnetic perturbations of the absorption type by the polar cap. In this period, errors up to 0.5 ph.c. are created in measuring the phase difference when signals pass through the polar regions. The seafarers are notified of the beginning of the absorption by the polar cap.

A significant improvement in the accuracy characteristics of the Omega RNS can be achieved by using a differential version of the system. Theoretical and experimental studies in the VLF range have shown the existence of a significant spatial and temporal correlation of the receiver phase readings located close enough to each other. The radius of this spatial correlation reaches 1,400 miles, the time correlation interval is several hours. Thus, the receiver phase readings, located within a radius of reliable spatial correlation—up to 300 miles, will contain approximately the same systematic random (absorption by the polar cap, sudden phase anomalies, etc.) errors.

Having established a reference point at a point with known coordinates in the area requiring navigation support of increased accuracy, continuous phase measurements are made in it with the signals of the RNS Omega on the standard receiver of the system.

The difference between the obtained phase count at each time and theoretically calculated one determines a correction that takes into account both random and systematic phase changes. These corrections are then transmitted via the communication channels to consumers within the range of the differential version of the Omega system, with a discreteness determined by the time correlation interval of phase changes in the system signals.

Transmitters of marine or aeronautical beacons can be used as transmitters for CPs. In some cases, the automatic introduction of translated corrections is used with the appropriate implementation of onboard receiver s and their retransmission equipment from CP.

Accuracy of position determination in the differential use of the Omega RNS rises to 0.3 miles at a distance of 50 miles, up to 0.5 miles at a distance of 100 … 200 miles and up to 1 mile at a distance of 300 miles from the CP, which is several times more accurate than the standard mode of operation of the RNS Omega.

The disadvantage of the differential method of Omega is that a significant number of CPs is required. For example, in order to have a $M \leq 0.25$ mile off the coast of the United States, 29 CPs are required. There are currently 15 differential regions in the world.

Quasi-differential use of the Omega RNS is possible. In accordance with this principle, at the time of obtaining high-precision observation in any way (for example, docking in a port, using a satellite navigation system, etc.), it is possible to calculate the value of the corrections and subsequently use them for a certain time interval in the definitions for RNS Omega.

The quasi-differential method is realized in practice in the form of the combined receiver s simultaneously receiving VLF RNS signals and satellite RNS signals. For example, in the MX-1105 receiver, differential corrections for the Omega RNS are determined from episodic observations of the satellite RNS. Amendments are used between satellite observations to determine the location with the RNS Omega quasi-differential method.

References

1. Bykov VI, Nikitenko YI (1976) Marine radio navigation devices. M Transp 399 pp
2. Yarlykov MS (1995) Statistical theory of radio navigation. M Radio commun 344 pp
3. Shatrakov YG (1999) Estimation of location error in the PI of the difference-ranging system and their dependence on the hyperbolic position lines. Col Ed (1). VIMI—C. 58–95
4. Olianyuk PV et al (1995) Radio navigation systems of the ultra-long-wavelength range. M Radio commun, 264 pp
5. Makarov GI et al (1973) Propagation of an electromagnetic pulse above the earth surface. Problems of diffraction and propagation of radio waves. Leningrad State University, 95 pp
6. Shatrakov YG et al (1991) Marine RNS. M Radio Commun 94 pp
7. Kinkulkin IE, others (1999) Phase method of coordinates determining. M Radio Commun 280 pp

Chapter 5
Instrumental Landing Systems

5.1 Systems of One Meter Range

In terms of aircraft landing capabilities in different weather conditions, the instrument landing systems are classified by the International Civil Aviation Organization (ICAO) [1–3]. The concept of the system's operational characteristics, which allow to provide landings with an established probability was adopted as the grounds for the classification.

Category I systems provide control of the aircraft to the decision-making height of 60 m with runway visibility range up to at least 800 m.

Category II systems provide control of the aircraft to the decision-making height of 30 m with runway visibility range up to at least 400 m.

Systems of the Category III are divided into three subgroups (IIIa, IIIb, and IIIc):

Category IIIa system provides aircraft control up to a decision height of 15 m with a visibility range up to a runway of at least 200 m;

System of the Category IIIb provides control of the aircraft to the surface of the runway in the absence of vertical visibility and along it;

System IIIc category provides control of the aircraft to the surface of the runway, along it, and during taxiing in the total absence of visibility.

The beacon antenna of the systems along the course path consists of two emitters, which form two cross-directed lines of the emission pattern. Through these patterns, the beacon emits in-phase amplitude-modulated oscillations with modulation frequencies of $F_1 = 90$ Hz and $F_2 = 150$ Hz. The field of these antennas can be represented in the form of

$$e_{1,2} = E_{m1,2} f_{1,2}(\varphi)\left[1 + m_{1,2}\sin\left(\Omega_{1,2}t\right)\sin\omega t\right] \quad (5.1)$$

where

$E_{m1,2}$ amplitude of field strengths in the maxima of the emission patterns;
$f_{1,2}(\varphi)$ normalized emission patterns;
$m_{1,2}$ coefficients of amplitude modulation.

Sauta O.I. et al., *Principles of Radio Navigation for Ground and Ship-Based Aircrafts*, Springer Aerospace Technology, https://doi.org/10.1007/978-981-13-8293-2_5

The resulting field with $E_{m1} = E_{m2} = E_m$ can be represented in the form of

$$\begin{gathered} e_p = e_1 + e_2 = E_m[f_1(\varphi) + f_2(\varphi)] \\ \{1 + [m_1 f_1(\varphi)\sin(\Omega_1 t)]/[f_1(\varphi) + f_2(\varphi)] \\ + [m_2 f_2(\varphi)\sin(\Omega_2 t)]/[f_1(\varphi) + f_2(\varphi)\} \sin \omega t] \end{gathered} \tag{5.2}$$

Coefficients of spatial modulation depth

$$M_1 = m_1 f_1(\varphi)/[f_1(\varphi) + f_2(\varphi)] \tag{5.3}$$

$$M_2 = m_2 f_2(\varphi)/[f_1(\varphi) + f_2(\varphi)] \tag{5.4}$$

The track line corresponds to the direction $\varphi = \varphi_0$ at which $M_2 = M_1$, and the difference in modulation depths of the MDD ($M_1 - M_2$) equals zero. The signal at the output of the receiver is proportional to the MDD, i.e., angular deviation from the runway axis.

Figure 5.1 shows the beam pattern of the beacon antenna and the spectra of the emitted signals.

The course beacons with a zero reference are also widely spread. Basically, they are used to create Category I systems. Here, the main (narrow) and additional (wide) channels are formed.

The glide path beacon with equal-signal emission has two antennas. The upper antenna emits amplitude-modulated oscillations with a modulation frequency of 150 Hz, the lower one −90 Hz.

If the phase shift of the currents feeding the antennas is zero, then the amplitude of the field strengths created by the upper and lower antennas in the far zone can be represented as

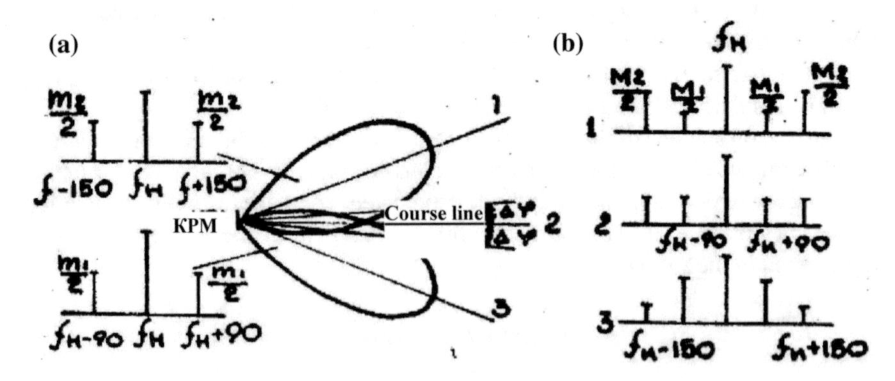

Fig. 5.1 Course path of landing systems: **a** patterns of the two antennas and the spectra of the emitted signals; **b** spectra of received signals according to the location of the mobile object

$$E_B = E_{mB} f_B(\Theta)(1 + m_2 \sin \Omega_2 t) \tag{5.5}$$

$$E_H = E_{mH} f_H(\Theta)(1 + m_1 \sin \Omega_1 t) \tag{5.6}$$

where

E_{mB}, E_{mH} amplitude of the field strengths of the upper and lower antennas;
$f_B(\Theta)$, $f_H(\Theta)$ normalized antenna patterns in the vertical plane;
m_1, m_2 coefficients of amplitude modulation.

The resulting field in the overlapping area of the emission patterns is as follows:

$$\begin{gathered} E_P = E_B + E_H = [E_{mB} f_B(\Theta) + E_{mH} f_H(\Theta)] \\ \{1 + [m_1 E_{mH} f_H(\Theta) \sin(\Omega_1 t)]/[E_{mB} f_B(\Theta) + E_{mH}(\Theta)] \\ + [m_2 E_{mB} f_B(\Theta) \sin(\Omega_2 t)]/[E_{mB} f_B(\Theta) + E_{mH} f_H(\Theta)]\} \end{gathered} \tag{5.7}$$

Coefficients of spatial modulation depth are

$$M_1 = m_1 E_{mH} f_H(\Theta)/[E_{mB} f_B(\Theta) + E_{mH} f_H(\Theta)] \tag{5.8}$$

$$M_2 = m_2 E_{mB} f_B(\Theta)/[E_{mB} f_B(\Theta) + E_{mH} f_H(\Theta)] \tag{5.9}$$

The deviation of the glide path indicator on the aircraft is in proportion to $M_1 - M_2$, i.e., MDD. On the glide path line (2°40′) MDD = 0.

5.2 Systems of the Decimeter Wave Range

ILS decimetre-waveband beacon system (landing beacon group—LBG) is intended to indicate landing course, the glide path, and the current range to touchdown to aircraft equipped with onboard equipment RSBN in meteorological conditions of ICAO Categories I and II.

The principle of the course-glide path beacons is based on the creation of equipotent zones in space using intersecting emission patterns. Each position of the emission patterns (one of the two lobes) at the moment of emission corresponds to a modulating repetition frequency of rectangular pulses ω_{M1}, ω_{M2} at a constant carrier frequency ω_H.

The work of the beacon is explained in Fig. 5.2, where its time patterns are shown. When processing position signals according to the given radio-sighting line onboard an aircraft, it is judged by the value of the radio-sighting coefficient (RBS).

$$KPC = \frac{U_1 - U_2}{U_1 + U_2}$$

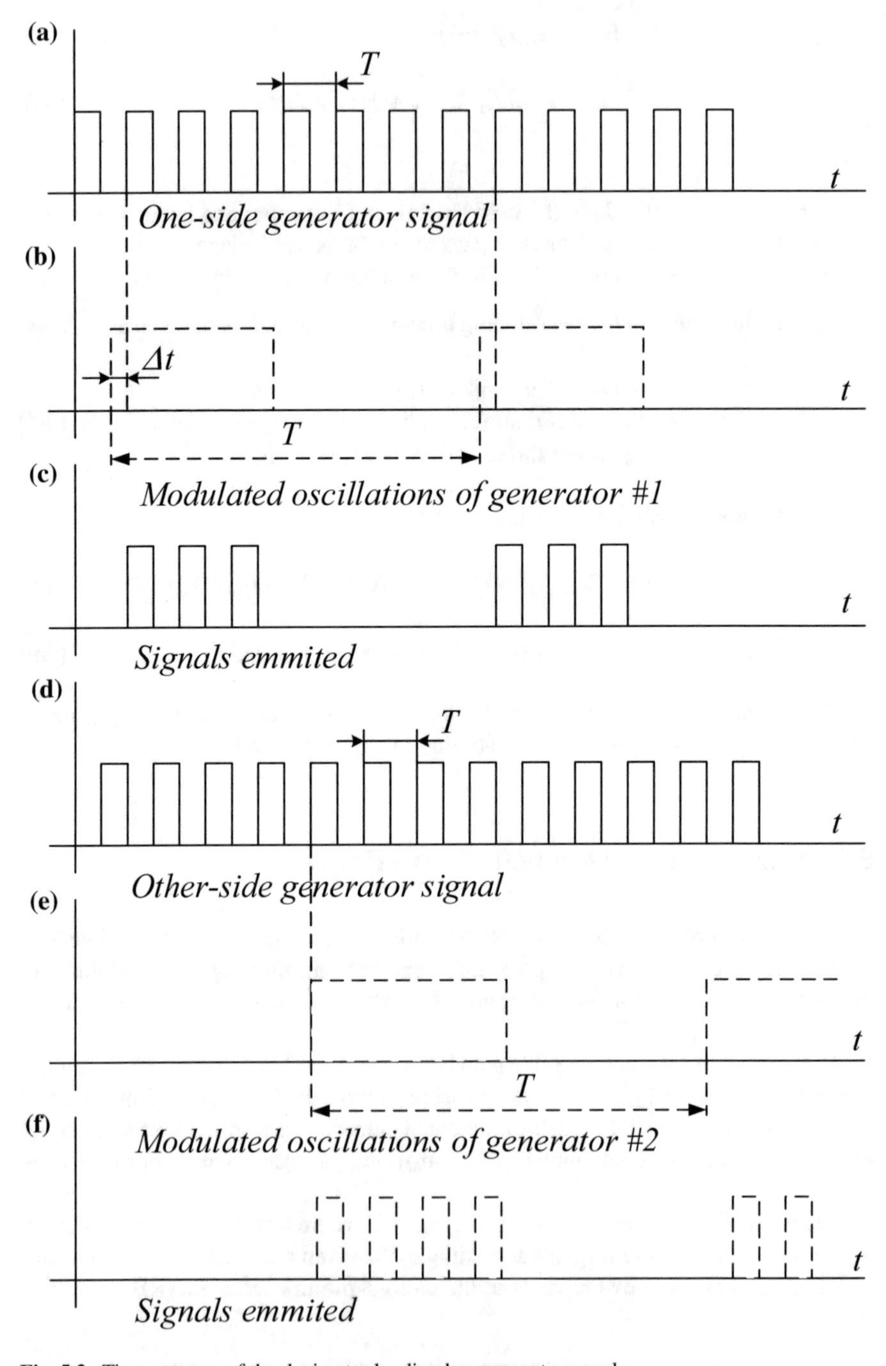

Fig. 5.2 Time patterns of the decimeter landing beacon system work

where U_1 is the amplitude of the main frequency voltage ω_{M1} in the signal spectrum (Fig. 5.2c); U_2—amplitude of main frequency voltage ω_{M2} in the signal spectrum (Fig. 5.2e).

Amplitudes U_1 and U_2 are extracted from the received signal by filters tuned to frequencies ω_{M1} and ω_{M2}, respectively.

Because the modulating signals are not synchronized, for each of them there is a time shift Δt relative to the commutation pulse origin. It is one of the sources of location errors according to the track line and the glide path.

In order to reduce the influence of the Earth on the formation of the equipotent signal zone in the glide path, antennas with narrow diagrams in the vertical plane are used for radio beacon.

Long-range repeater includes a receiving and transmitting device. The latter consists of radio-frequency receiver units and a control unit and is designed to convert the received request signals into high-frequency coded packets.

5.3 Microwave Systems

The system provides information about the aircraft over the angle in the azimuthal plane relative to the axis of the runway, elevation angle, and range. The system operates in a six-centimeter wavelength range and provides landing of aircraft under meteorological minimums of Categories I, II, and III. Determining of the angular position of the aircraft in the system is based on the onboard equipment measurement of the time interval $t_{\varphi,\Theta}$ between two pulses I-A and I-B, appearing at the output of the onboard receiver when the aircraft is irradiated by the beacon antenna during the forward and reverse travel of the beam of this antenna. The angles in the azimuth and angle planes are determined by taking

$$\varphi = M_{\varphi}\left(t_{\varphi} - T_{0\varphi}\right) \tag{5.10}$$

$$\Theta = M_{\Theta}(t_{\Theta} - T_{0\Theta}) \tag{5.11}$$

where

$T_{0\varphi}$, $T_{o\Theta}$ the time intervals between pulses I-A and I-B when the aircraft is on the landing course and the glide path of descent;

M_{φ}, M_{Θ} scale factors equal to half the scanning speed of the antenna beam of the course and glide path beacons.

The time counting in the onboard device is accrued from the moment of reception of the starting pulse I-0. The starting pulse is transmitted by the antenna during the emission of the preamble preceding the beginning of the beam scan in the forward direction.

Onboard, the moment $t_ц$ corresponding to the middle of the beam scan cycle is also noted.

The time interval $T_ц$ between the I-0 and the moment $t_ц$ depends on the angular position of the aircraft.

Figure 5.3 gives the necessary explanations on the principle of the system's operation.

The use of the microwave landing system allows, due to the use of a narrow beam in the directivity pattern (1°), to exclude the effect of local objects and Earth on the characteristics of the system. Depending on the type of aircraft, the crew can choose the optimal glide path.

The system includes course, glide path, and long-range (DME beacon) beacons.

For Category III systems, the reverse-mode beacon is also used, which makes it possible to solve the problem of missed approach in case of landing failure.

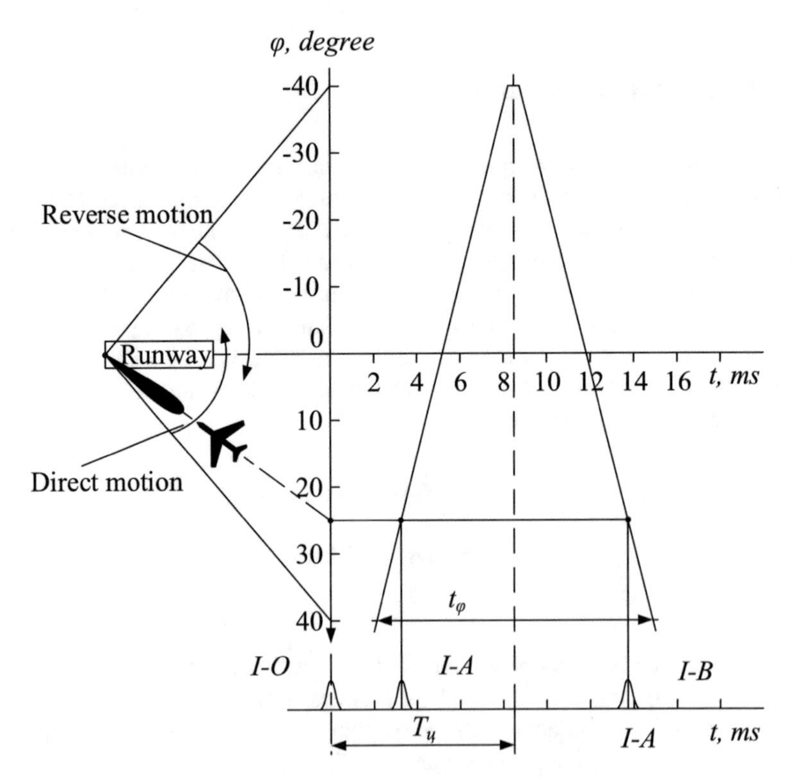

Fig. 5.3 To explain the operation principle of the microwave landing system

5.4 Prospects for the Development of Instrumental Landing Systems

The development of space-based radio navigation systems GLONASS and GPS dictates a fundamentally new approach to the creation of instrumental landing systems for all types of aircraft, as well as other automated systems associated with traffic control. SRNS GLONASS and GPS allow to obtain geographical coordinates, altitude, speed, and acceleration on mobile objects. This information is sufficient to generate control signals for the onboard equipment to indicate the required position of the aircraft on the descent path to the runway, taking into account the distance to the touchdown. The transfer of additional information to the aircraft allows the system to work in the relative navigation mode and to increase the accuracy of the aircraft fixation on the descent path to the runway. Tests of equipment in this mode of operation continue from the moment the AS grouping in the systems is elevated to the standard mode. The results obtained by the leading US company Raytheon allowed to plan the introduction of the JPALS system (Joint Precision Approach and Landing System) for all airfields and the Navy aircraft carriers. The introduction of JPALS to replace the ILS and MLS systems is scheduled from 2014 to 2021. One of the reasons for replacing the traditionally used instrument landing systems of aircraft with JPALS is the high operating costs of the former. The Raytheon company that developed and implemented the JPALS system, has an annual turnover of 25 billion US dollars. It is the leader in the production of high-tech digital avionics. Since the company was established in 1921 with a small group of several people, it grew to a current staff of more than 75 thousand people.

References

1. ICAO Annex 10 Volume 1 Aeronautical Telecommunications—Radio Navigation Aids
2. Sauta OI et al, under editorship of Shatrakov YG (2016) Development of navigation technologies for improving flight safety, SPb, GUAP, 299 pp
3. Yarlykov MS (1995) Statistical theory of radio navigation. M. Radio and Communication, 344 pp

Chapter 6
Satellite Radio Navigation Systems

6.1 Overview of the Development of the Systems

The launch of the world's first Soviet artificial earth satellite (AES) took place on October 4, 1957.

In 1959, the first US navigational satellite was launched into orbit, and in 1964 a low-orbit transit system was put into operation to aid the American nuclear-powered Polaris submarines.

In 1967, the first navigation satellite Kosmos-192 was launched into orbit in the USSR in order to create a low-orbit system of the SNS Cyclone. The complete system was put into operation in 1976 comprising of six spacecraft. In 1976, a civilian version of the navigation system was developed for the needs of the merchant navy, known as the Cicada.

The first US GPS satellite was launched into orbit on February 22, 1979. The creation of the GPS constellation was completed in 1989.

The first GLONASS satellite was launched in the USSR on October 12, 1982. The GLONASS system was officially launched on September 24, 1993. In January 1994, the system consisted of 24 satellites. In the period 1995–2004, the number decreased to 12. In 2004, the rebuilding of the group began.

Galileo is the European SNS project. The European system is designed to provide navigation for any moving objects with less than 1 m accuracy. It is expected that Galileo will enter service in 2019 when all 30 of the planned satellites (27 operational and 3 reserve ones) will be brought into orbit. The space segment will be complemented by ground infrastructure, which includes two control centers and a global network of transmitting and receiving stations. The first prototype satellite of the Galileo system was launched on December 28, 2005, into an estimated orbit with an altitude of more than 23 000 km with an inclination of 56°.

No state is willing to depend on its development in any way on another state, albeit friendly at the moment. Therefore, the search for an alternative to GPS and GLONASS led to the creation of GALILEO.

Sauta O.I. et al., *Principles of Radio Navigation for Ground and Ship-Based Aircrafts*, Springer Aerospace Technology, https://doi.org/10.1007/978-981-13-8293-2_6

On April 14, 2007, the first launch of the Chinese navigation satellite Beidou (Great Bear) was held to create the Compass SNS. On April 14, 2009, the second navigation satellite was launched into geostationary orbit. The GNSS subsystem developed by China was intended for use only in that country. The creation of the space navigation system of the PRC was to be completed by 2015. The space segment of the SNS will be formed from 5 satellites in the geostationary orbit and 30 satellites in the medium earth orbit. The system will be fully compatible with Russian GLONASS, European Galileo, and American GPS.

Indian Regional Navigation Satellite System, abbreviated as IRNSS. The first satellite was launched in 2008. The satellite group IRNSS will consist of seven satellites in geosynchronous orbits. In that four satellites out of seven in the IRNSS will be placed in orbit with an inclination of 29° to the equatorial plane. The project completion date set was for 2011, IRNSS will provide only regional coverage in India and parts of neighboring states.

Japanese Quasi-Zenith navigation system (QZSS). A satellite has been launched, only three satellites will be included in the satellite segment, the orbits of which will be selected in such a way that their subsatellite points follow the same trajectory on the Earth's surface with the same time intervals. At the same time, at least one satellite will be visible at an elevation angle of more than 70° at any given time in Japan and Korea. This feature also defines the name of the navigation system—Quasi-Zenith (Quasi-zenith). This feature is important for mountainous areas or cities with tall buildings.

In the near future, three global navigation satellite systems, GPS, GLONASS, and GALILEO, will work simultaneously. Almost all countries currently use GPS only, the normal operation of which depends entirely on the US government.

The use of satellite navigation systems (SNS) for navigation purposes presents significant advantages over conventional navigation aids.

The SNS are characterized by higher accuracy than the systems currently in operation. Combined with the functional additions and with the air-ground data transmission systems, the SNS allow automatic dependent surveillance (ADS) in any area of the airspace. The introduction of the SNS and the possible decommissioning of ground-based navigation aids will significantly improve the regularity, efficiency, economy, and safety of air transport operations.

Due to the fact that GPS and GLONASS are used in practical activities, they will be considered in detail later.

6.2 Principles of System Operations

The principle of determining the customer's coordinates in the SNS is based on measuring distances to navigation satellites [1–3].

The geometric interpretation of this principle realization can be explained as follows. Suppose that at any given moment in time, the positions of satellites in near-Earth space are known and the primary navigation parameters can be measured—the distances to satellites in the field of view of the SNS receiver. The measured distance D_1 to one satellite determines the position surface in the form of a sphere with a radius equal to the measured distance (Fig. 6.1).

The distances D_1 and D_2 to two satellites define two position surfaces, the intersection of which determines the line of position in the form of a circle. The position surface obtained with a third satellite in the form of a sphere with radius D may intersect with the position line in the form of a circle obtained from the first two satellites, only at two points M_1 and M_2. Thus, the measured distances to three satellites limit the possible position to two possible points. The method of logical exclusion determines which of the two points is the position of the SNS receiver. For example, if one of the points is too far from the surface of the Earth, or has too high a velocity relative to the Earth's surface, or is very far from a previously determined position, then such a point cannot be the desired position. In the onboard equipment computers, there are several algorithms that make it possible to distinguish between the correct position and the false one.

Determination of the distance D from the satellite to the SNS receiver is performed as a result of measuring the time in which the radio signal passes from the satellite to the SNS signal consumer by the formula:

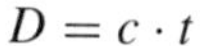

$$D = c \cdot t$$

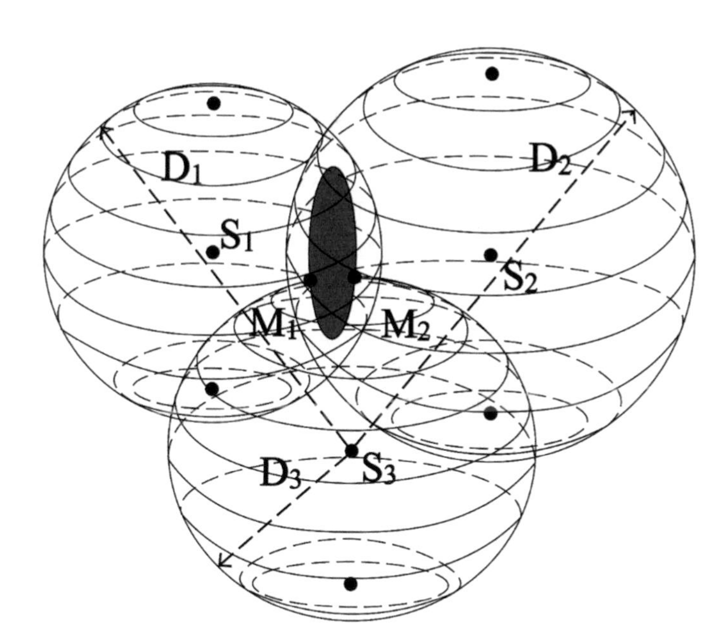

Fig. 6.1 Determining position with navigation satellites

where c is the speed of radio signal propagation; t is the time of the radio signal passing.

For distances close to $D = 21\ 000$ km and speed $c = 300\ 000$ km/s, the time of the signal passage $t = 0.07$ s. Therefore, high-precision measurement of very short time intervals needs to be realized in the consumer's equipment.

To determine the time of the radio signal passing from the satellite to the signal receiver, a method of comparing the pseudorandom codes generated in the equipment of the satellite and the SNS receiver was used.

In its most general terms, this method consists of the following. The equipment of satellites and receivers is synchronized with very high accuracy. Both on satellites and in receivers, the same sequences of very complex digital codes are generated simultaneously. These codes are so complex that they seem to look like long chains of random pulses, which are commonly called pseudorandom codes. So, since the equipment of satellites and receivers generates the same codes at the same time, the time of the signal passing from the satellite to the receiver is determined by the delay of the received code. Generated pseudorandom codes are repeated every microsecond, i.e., every 10^{-6} s.

Most SNS receivers provide a time measurement to within $\Delta t = 10^{-9}$ s (i.e., up to 1 ns).

To determine the position of the consumer accurately, it is necessary that the accuracy of clock synchronization on satellites and in consumer equipment conforms to the required accuracy of measuring the time of the radio signal passing from the satellite to the receiver.

The satellites are equipped with sets of four high-precision atomic clocks and, in addition, these clocks are corrected by ground control stations. In the SNS receivers, a relatively inaccurate quartz clock is installed.

The error in determining the moment of time Δt by the clock of the receiver in comparison with the reference at the satellite clock is determined by the computer of the consumer equipment as a result of special algorithm calculation.

Suppose that the satellites and the SNS receiver are in the same plane.

If $\Delta t = 0$, i.e., there are no errors in measuring the time of a radio signal passing simultaneously from three satellites, then the position lines will intersect at one point. In case when the error $\Delta t \neq 0$, the calculated position lines will be separated from the actual position lines by a value of $c \cdot \Delta t$ and form a certain range of possible positions of the SNS receiver (error range). The dimensions of this range are determined by the value $c \cdot \Delta t$ and angles of the intersection of the position lines. According to a special algorithm in the computer of the SNS receiver a value Δt is calculated after a series of measurements, which becomes the third coordinate determining the position of the receiver on the plane.

To calculate the error in determining the time by the clock of the SNS receiver and the location of the SNS receiver in space (i.e., in a system of three coordinates), simultaneous measurement of the distances to four satellites is necessary. In this case, the error in time Δt is a fourth coordinate, and therefore the four position surfaces in the form of spheres with radii equal to the corresponding distances from the four satellites determine a certain region of possible positions of the SNS receiver.

Thus, for real-time high-precision position determination, a combination of a multichannel receiver and a high-speed computer is necessary. The receiver provides simultaneous reception of signals from four satellites, the computer calculates the error of the clock At and the receiver coordinates in the selected coordinate system.

Where it is possible to receive signals from only three satellites, the algorithms of the SNS consumer equipment assume the Earth's center to be the fourth satellite, therefore, one sphere of position is a sphere with a radius equal to the distance from the center of the Earth to the consumer (to the SNS receiver). The distance from the center of the Earth to the surface of the general Earth ellipsoid is calculated by the computer of the SNS consumer equipment, and the distance from the surface of the ellipsoid to the AC (absolute altitude) is entered into the computer automatically.

High-precision positioning is possible only if the coordinates of the navigation satellites are accurately calculated at the time of measuring the distances to the satellites. The working orbits of the satellites are chosen in a way to ensure the high accuracy of the satellites' maintaining the specified orbits and the period of revolution relative to the center of the Earth. The ephemeris of the satellites (the parameters determining their position in the orbit) is determined and refined by means of a ground command-measuring complex. Information on the ephemeris of all satellites in the form of a so-called almanac is entered into the computer's memory of the SNS signals consumer equipment. Consequently, the computer is provided with data for calculating the coordinates of the satellites at any given time. Although, since under the influence of gravitational pulsations of the Moon and the Sun and the pressure of solar radiation on the surface of the satellite, there are changes in the ephemeris of the satellite, then on the ground command-measuring complex ephemeris errors are determined, which are transmitted to the satellite. The signals transmitted by the satellite contain information about the ephemeris errors of this satellite.

Data on ephemeris of the satellite contained in the almanac and information on ephemeris errors provide a highly accurate calculation of satellite coordinates by the SNS receiver computer.

The possible accuracy of measuring the distance to the satellite is estimated by the total mean square error in determining the distance to the satellite $\sigma_d = 5{-}10$ m. The accuracy of determining the position of the SNS receiver is determined not only by the errors in the measurement of distances to the satellites, but also by the relative location of the surfaces of the SNS receiver position, i.e., the relative position of the satellites.

Suppose that two satellites and an SNS receiver are located in the same plane. Both position lines are determined with an AD error. Then the region of the possible position of the SNS receiver will be much smaller with the angle of intersection of the position lines close to 90° (Fig. 6.2a) than with the angle of intersection close to 180° (Fig. 6.2b). The possible error in determining the coordinates due to the so-called geometric factor can increase several times.

When measuring distances of to four satellites, the AD errors also determine the area where the SNS receiver may be located, and the accuracy of determining the coordinates is highly dependent on the relative position of the satellites. If there are more than four satellites in the field of view of the SNS receiver antenna, then

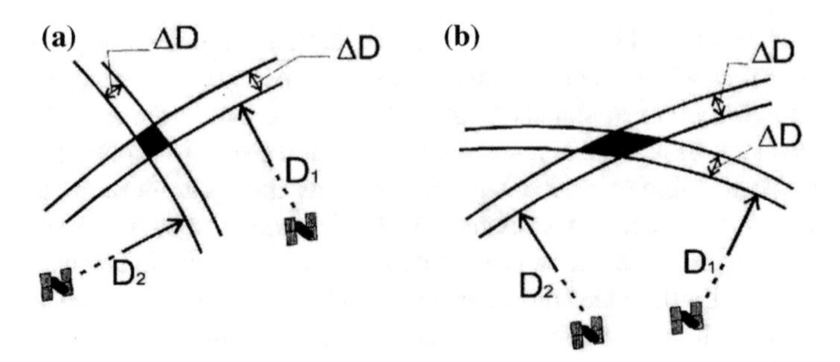

Fig. 6.2 The influence of the geometric factor on the position determination accuracy

according to a certain algorithm, four satellites can be selected, whose relative position provides the highest accuracy of computing the position of the consumer at the moment.

To assess the effect of the relative positioning of satellites and the SNS receiver on the accuracy of determining the coordinates, the criterion GDOP—Geometric Dilution of Precision is used (Geometric accuracy decrease, geometric factor). The geometric factor GDOP is, as a rule, denoted by the abbreviation DOP (in some SNS receivers this criterion is denoted by PDOP—Position Dilution of Precision).

Geometric Dilution of Precision (GDOP) is the relative value of a position determination error using a system that forms several families of lateral position surfaces. More precisely, this is the ratio of the root mean square error of position determination to the root mean square value of measurement errors, assuming that all components of the measurement error are statistically independent, have zero expected value and the same standard distribution. From the observer's point of view, GDOP is an indicator of the perfection of the geometric characteristics of the signal sources location that form these lateral position surfaces. A low GDOP value is desirable, whereas a high is not.

Due to the optimal choice for bearing finding four satellites from the antennas, which are in the field of view, usually six to eight satellites, the accuracy of determining the position of the consumer is increased 4–6 times.

Thus, high accuracy in determining the position of the SNS receiver is provided based on the following basic principles:

- using the distance to the satellite as the primary definable navigation parameter;
- determining the time of the signal passing from the satellite by means of a special pseudo-code generated on satellites and in the consumer's equipment;
- ensuring accurate clock synchronization on satellites and in consumer equipment;
- calculations in the consumer equipment using an almanac and ephemeris corrections of the satellite coordinates;
- optimal selection of satellites for bearing finding taking into account their relative location.

It is very important to bear in mind that in order to determine the location it is desirable to receive signals from those satellites that are above the horizon. In this connection, there is the concept of angle mask.

The angle mask. Fixed elevation angle relative to the user's horizon, the satellites below which are ignored by the receiver software. Mask angles are used mainly in the analysis of GNSS characteristics and are used in some receiver models. The angle mask is determined by the characteristics of the receiver, the power of the transmitted signal at low elevations, the sensitivity of the receiver, and errors acceptable for small elevations.

The minimum applicable mask angle.

The minimum elevation angle of the satellite relative to the user's horizon, at which the satellite can be reliably used for navigational calculations. The change in the minimum applied mask angle depends on the environment, the design and location of the antenna, the height and angular position of the user.

The use of pseudorandom code in the SNS is due not only to the need to provide a high-precision measurement of the time of a radio signal passing from the satellite to the SNS receiver, but also the need to receive and process very weak radio signals from the satellite.

The signals from the navigation satellites are so weak that they cannot be registered against the Earth's natural radio emission background. The Earth's natural radio noise is a random variation of electronic pulsations. While the received pseudorandom code is a strictly defined sequence of electronic pulses. At that, since the pseudorandom code is repeated every microsecond, it is possible for a high-speed computer to perform multiple comparisons of the received signals and to allocate a pseudorandom code against the Earth's natural radio noise background. As a result, the SNS receiver can have a very small antenna, and in general, the consumer equipment can be relatively small in size and weight and, in addition, of a relatively small cost. This, in turn, facilitates the conversion of the SNS into a mass-use system.

One of the most important reasons for the use of pseudorandom code in the SNS is the expediency of using all satellites of the same carrier frequency in their transmitters. Although, since each satellite transmits only its inherent code, the receiver can easily distinguish the signals of a particular satellite, and the satellites do not jam each other, working on the same frequency.

The use of pseudorandom code in the SNS also allows the system owner to control the mode of access to the system.

So in American GPS, two types of codes are used:

C/A-code (Clear/Acquisition-code—code of free use);
P-code (Protected-code).

P-code is classified and only the US Department of Defense has access to it. C/A-code is public. However, using C/A-code, the S/A (Selective Availability-limited access) mode can be introduced, which is designed to reduce the accuracy of bearing finding by simply coarsening the time signals transmitted by the satellite.

To improve the accuracy of determining the customer's coordinates, for example, during the approach, a differential mode is provided in the SNS.

The navigation satellite rotates relative to the Earth's center of mass in a practically circular orbit with an altitude above the Earth's surface of about 20 000 km and its position can be calculated in a coordinate system centered at the center of the Earth. The position of the SNS signals consumer needs to be calculated relative to the Earth's surface in the accepted geodetic coordinate system.

The position of the navigation satellite relative to the Earth's center of mass is determined in the rectangular absolute geocentric coordinate system OXYZ, connected with the current equator and the vernal equinox.

In the GLONASS system, the law of motion of the *i*th navigation satellite is given by: rectangular coordinates x_i, y_i, z_i and the rate of their change x_i', y_i', z_i' in respect with a certain definite time t_0. If the forces acting on the satellite are known, then the equations of satellite's motion can be made. Consequently, for any instant of time t, the coordinates of the *i*th satellite x_i, y_i, z_i and the rate of their change x_i', y_i', z_i' can be determined by integrating the satellite motion equations under certain initial conditions. As a result of determining the distance D_1 to the *i*th satellite, the following equation can be made (Fig. 6.3):

$$(x - x_i)^2 + (y - y_i)^2 + (z - z_i)^2 = D_i^2 \tag{6.1}$$

where

x, y, z coordinates of the SNS receiver relative to the Earth's center of mass;
x_i, y_i, z_i coordinates of the *i*th satellite relative to the Earth's center of mass.

Assuming that the error in measuring the time of a signal passing from a satellite to a SNS receiver $\Delta t = 0$, based on the results of receiving signals from three satellites, a system of three equations can be made:

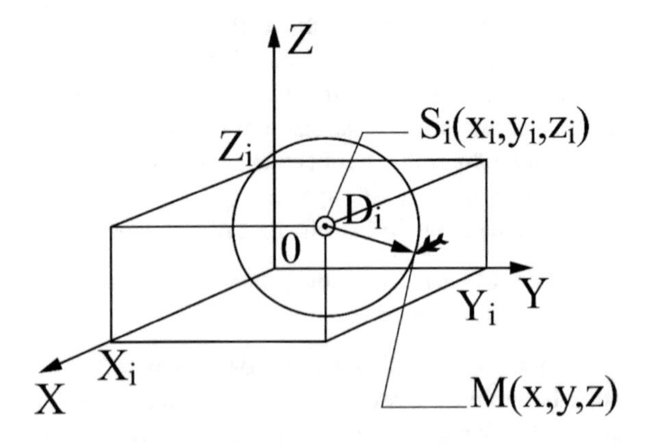

Fig. 6.3 Geocentric coordinates of the satellite and aircraft

$$\left.\begin{aligned}(x-x_1)^2+(y-y_1)^2+(z-z_1)^2&=D_1^2\\(x-x_2)^2+(y-y_2)^2+(z-z_2)^2&=D_2^2\\(x-x_3)^2+(y-y_3)^2+(z-z_3)^2&=D_3^2\end{aligned}\right\} \quad (6.2)$$

and the coordinates x, y, z of the consumer of the SNS signals relative to the Earth's center of mass are determined. Then geocentric rectangular coordinates are converted into geodesic ones.

The relationship of coordinates x, y, z with latitude B, longitude L and height H can be given by the following system of equations:

$$\left.\begin{aligned}x&=(N+H)\cos B\cos L\\y&=(N+H)\cos B\sin L\\z&=[(1-e^2)N+H]\sin B\end{aligned}\right\} \quad (6.3)$$

where

x, y, z are rectangular coordinates;
B, L, H geodetic coordinates (respectively, latitude and longitude, rad, and height, m);
N is the radius of curvature of the first vertical, m;
e is the eccentricity of the ellipsoid.

The values of the curvature radius of the first vertical and the square of the eccentricity of the ellipsoid are calculated according to the formulas:

$$N=\frac{a}{\sqrt{1-e^2\sin^2 B}} \quad (6.4)$$

$$e^2=2\alpha-\alpha^2 \quad (6.5)$$

where

a the semimajor axis of the ellipsoid, m;
α compression of the ellipsoid.

Equations (6.2) and (6.3) characterize the algorithms solved in the computers of onboard SNS equipment only in general terms. In addition, in order to determine the coordinates of the SNS receiver, an error in the measurement of the signal transit time Δt from the satellites to the SNS receiver needs to be calculated.

When determining the user's location using GPS, the problem is solved in the 1984 world geodetic coordinate system (WGS-84—World Geodetic System 1984), and when using GLONASS—in the 1990 system of geodetic coordinates Earth parameters, version of the year 2000.

When performing an accurate approach (with vertical guidance) using the SNS, it is not necessary to take into account the geoid wave for the runway threshold because the altitude of the aircraft is determined in the WGS-84 or EP-90.11 coordinate system, and the aircraft is landing on the runway, the elevation of which is measured in the same coordinate system.

Aeronautical satellite navigation systems provide not only a way to determine the position of the aircraft, but also to determine the track angle and the ground speed of the aircraft. To determine the track angle and the ground speed, shift of frequencies is measured in signals from each satellite.

A complete algorithm for determining the coordinates, ground speed, and track angle of an aircraft (SNS signal consumer) provides

- calculation with the help of the almanac and ephemeris corrections of the calculated coordinates values of each of the four satellites: x_i, y_i, z_i and the rates of change of these coordinates: x_i', y_i', z_i';
- measuring the time of a signal passing t_i from satellites;
- calculation of errors Δt_i in measuring the time of signals passing from satellites;
- calculation of distances R_i to satellites;
- calculation of the geodetic coordinates of the latitude φ and longitude λ of the aircraft, as well as the ground speed and the track angle of the aircraft.

When only three satellites are locked by the consumer equipment and, therefore, the elevation of the SNS receiver relative to the mean sea level is entered manually or automatically (in navigation complexes), the radial error in the aircraft position can be twice the elevation error. So, for example, if the absolute height is entered with an error of $\Delta_h = 100$ m, then the radial error in the position of the aircraft can reach a value of $\sigma_r = 200$ m.

The actual track angle and track speed can be determined with high accuracy only when the speed of the SNS receiver (aircraft speed) exceeds 30 knots (55.6 km/h). In this case, the mean square error in determining the actual ground speed, approximately, $\sigma_w = 0.1$ knots (0.2 km/h).

One of the main ways of using artificial Earth satellites in aviation is their use for aircraft navigation (aircraft). The SNS has a number of advantages over traditional radio engineering systems (RTS) navigation:

- the high altitude of the satellite makes it possible to create a global coverage zone for radio technical means installed on satellites, using rather simple antenna devices, both on the satellite and on the aircraft;
- with the help of the satellite constellation, the creation of a navigation system covering the territory of the globe;
- the finding of a satellite within the line of sight at any point in the range of its radio engineering means allows using the most noise-resistant ranges of radio waves and transmitting signals with the least distortion;
- almost unlimited bandwidth of the SNS;
- relative simplicity and cheapness of onboard SNS equipment due to the lack of a transmitter and modern signal processing technologies;
- with the further development of the SNS the integrated use of satellite systems is possible to solve problems of navigation, communications, and surveillance.

Navigation satellites in modern SNS used for navigation are located almost in circular orbits with an altitude of about 20 000 km. It is established that when the satellites move at such altitudes, the forces acting on the satellites and creating deviations from the calculated orbits are highly stable, which allows to predict accurately the movement of the satellite for several months in advance.

The noted advantages of the SNS allow, with their implementation, to significantly promote the solution of a number of tasks to provide air traffic services. The most important of them are

- increasing the flight safety levels;
- increasing the accuracy and integrity of navigation, especially in areas with an underdeveloped infrastructure of ground-based equipment of navigation RTS and over water;
- reducing aircraft separation intervals and increasing airspace capacity;
- straightening of air routes.

In addition to indirect benefits, due to the increase in the accuracy and reliability of air navigation, the direct costs of States providing air traffic are significantly reduced.

In view of the undeniable technical and economic advantages of the ICAO satellite systems, it was decided to establish a worldwide satellite communications, navigation, surveillance, and air traffic management system—CNS/ATM (Communication, Navigation, and Surveillance/Air Traffic Management) using the GNSS Global Navigation Satellite System based on GPS and GLONASS. A specially created ICAO committee developed a concept for future CNS/ATM air navigation systems. In terms of the scale of change, the transition to CNS/ATM systems is the largest program the aviation community has ever had to deal with.

6.3 Main Characteristics of Systems

GPS and GLONASS are autonomous medium-orbit satellite systems that are designed to determine spatial coordinates of mobile and stationary objects on the Earth surface and in near-Earth space, with high accuracy, and make precise coordination of time.

The operation principle of both systems is similar: both GPS and GLONASS consist of the following three main segments:

- Spacecraft subsystems;
- Monitoring and control subsystems;
- Consumers' navigation equipment.

The spacecraft subsystem of the GLONASS system consists of 24 satellites moving in circular orbits at 19 100 km height, inclined to 64.8°, with an orbital period of 11 h 15 min in three orbital planes. Orbital planes are separated longitudinally by 120°. In each orbital plane, there are 8 satellites arranged with a uniform shift in a latitude argument of 45°. In addition, the planes themselves are shifted relative to one another by a latitude argument of 15°. Such a configuration of the Spacecraft Subsystems makes it possible to provide continuous and global coverage of the Earth's surface and near-Earth space by a navigation field.

The monitoring and control subsystem consists of the GLONASS System Control Center and a network of measurement, control, and monitoring stations spread throughout Russia. The Monitoring and Control Subsystems perform monitoring the correct functioning of the Spacecraft Subsystems, continuously refining the orbit parameters and sending temporary programs, control commands and navigation information to satellites.

The consumers' navigation equipment consists of navigation receivers and processing devices used to receive navigation signals.

The navigation equipment of the GLONASS system users performs unconditional measurements of pseudo-range and radial pseudo-speed for up to four (or three) GLONASS satellites, as well as reception and processing of navigation messages contained in satellite navigation signals. The navigation message describes the satellite position in space and time. As a result of processing the received measurements and the received navigation messages, there determined three (or two) consumer's coordinates, three (or two) components of the speed vector of its movement, and the consumer's time scale is linked to the Universal Time Coordinated (UTC) scale.

Data that provide planning of navigational definitions sessions, selection of a working "constellation" of navigational spacecraft and detection of radio signals transmitted by them, are transmitted as part of the navigation message.

In the GPS system, satellites are evenly arranged into 6 orbits which planes are inclined to 55° to the equatorial plane, and there are 4 satellites on each orbit. Orbits are spaced along the equator at 60° interval.

The main technical data of GPS and GLONASS are shown in Table 6.1.

For civil GPS users, standard positioning using course code in selective access (S/A) mode or without this mode is provided. For a more accurate location, there used a fine code with the approval of the US Department of Defense (see Table 6.1).

The of GNSS operating principles are relatively simple, but for their implementation, the advanced achievements of science and technology are used.

All GPS or GLONASS satellites are equitable in their system. Each satellite, through a transmit antenna, emits a coded signal at two carrier frequencies (*L1*; *L2*) which can be received by the corresponding user's receiver located in the satellite's coverage area. The transmitted signal contains the following information:

- ephemeris of satellites;
- ionospheric modeling coefficients;
- information on the state of the satellite;
- system time and maintenance of satellite clock;
- information on the satellite's drift.

The receiver of the airborne equipment of the aircraft generates a code identical to that being received from the satellite. When comparing two codes, a time shift is determined that is proportional to the distance to the satellite. When receiving simultaneous signals from several satellites, it is possible to determine the receiver

Table 6.1 Main characteristics of GPS and GLONASS

Parameters	GPS	GLONASS
Satellites		
Number of satellites	32[a]	24
Number of orbits	6	3
Orbits height, km	20 200	19 100
Sidereal period, h:min	11:56	11:15
Orbit inclination, deg	55	64.8
Carrier frequency (L1), MHz	1575.42	1596…1605
Power source	Solar battery and battery	
Ground stations		
Main control station	1	1
Monitoring stations	5	2
Loading stations	–	4
Laser tracking stations	–	1
Accuracy of determination (95%)		
Position in plan, m	5	7
Altitude, m	8	12
Speed, m/s	0.03	0.3
Time, ns	3	5
Coverage area	Global	
Number of concurrent users	Unlimited	

[a]Actual number of satellites (standard is 24). Actual number of satellites can vary

location with high accuracy. Obviously, for the functioning of the system, precise synchronization of the codes generated by satellites and receivers is necessary.

The key factor determining the system accuracy is that all components of the satellite signal are precisely controlled by atomic clocks. Each satellite has four quantum generators, which are high-precision cesium standards of frequency, and their daily instability is 5×10^{-13}. The accuracy of the mutual synchronization of the onboard satellite timescales is 20 ns. The basis for the system timescale formation in the SNS is the hydrogen standard for the frequency of the system central synchronizer, and its daily instability is 5×10^{-13}. The clock of the receiver is less precise, but its code is constantly compared with the satellite clock and corrected to compensate the clock inaccuracy.

The ground segment monitors the satellites, performs control functions, and determines navigation parameters of the satellites. Data on the measurements results performed by each monitoring station are processed at the main control station and used to predict the satellite ephemeris. The main control station generates signals to correct the satellite clock.

The position of the aircraft using GPS and GLONASS is determined in geodetic coordinate systems.

Depending on the purpose of the reception indicators (RI), SNS are divided into three groups:

geodesic;
navigation;
domestic (Fig. 6.4).

The following RIs used only for air navigation are described further.

The satellite navigation signal receiver (GLONACC/GPS-receiver) is a microcircuit or a set of microcircuits (sometimes in context with or without an antenna) with the appropriate software to receive and decode the satellite navigation system signals and output the object coordinates in a certain format. If the receiver transmits information to the navigation complex, it is an SNS sensor.

An alternative method to check the reliability of the information received from the SNS is to compare this information with navigation information obtained from other navigation systems, such as INS, LORAN-C, and VOR/DME. This method is implemented in hardware and called AAIM. This method has the advantage over RAIM—there is no need to process signals from an additional satellite, and this allows to continue navigational determinations with guaranteed reliability in the visibility of only 4 satellites, as well as an anomaly effect to the satellite group (auroral storm).

Use of the information on the barometric altitude when linking SNS equipment with an altitude sensor is performed in order to

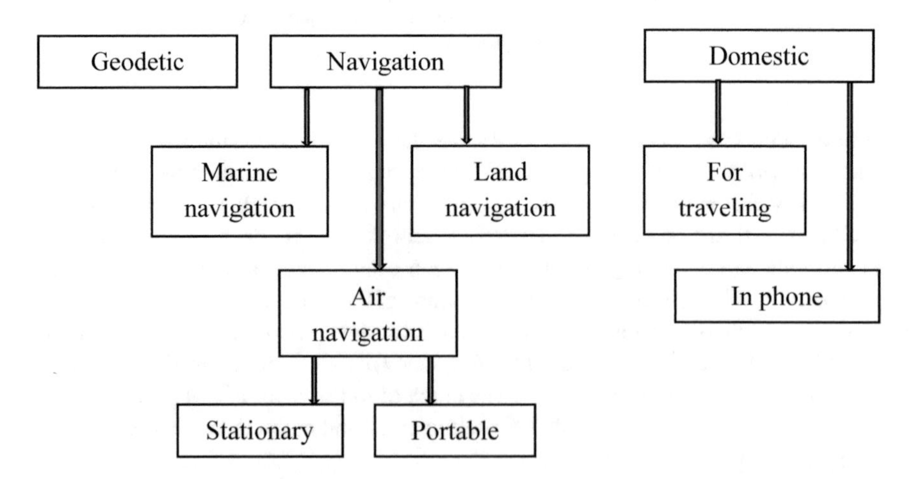

Fig. 6.4 Classes of SNS receivers

"Harmonization" of navigation definitions, which will significantly speed up the procedures of mathematical filtration;
"Support" RAIM, when the barometric altitude is used as a position sphere only for the RAIM algorithm, and only in cases where it is impossible to carry out RAIM with the fifth satellite (that is, signals are processed only from 4 satellites);
"Support" navigation definitions, when the barometric altitude is used as a position sphere in the "Approach" mode (when QNH input) and provided that there are not enough visible satellites to operate the equipment in the "3D" mode, i.e., when 3 satellites are visible.

References

1. ICAO annex 10 volume 1 aeronautical telecommunications—radio navigation aids
2. Olianyuk PV et al (2008) Satellite navigation systems. Spb. AGA, 98 pp
3. Sauta OI et al, under editorship of Shatrakov YG (2016) Development of navigation technologies for improving flight safety. SPb, GUAP, 299 pp

Chapter 7
Differential and Relative Operation Modes of Systems

7.1 Principle of Systems Operation in Differential Mode

The operational experience of the SNS shows that the signals emitted by the navigation satellites are subject to various interferences: unintentional and deliberate interference, as well as atmospheric interference [1–9].

Unintentional interference. Most of the interference effects on the SNS are related to inboard systems, and as a result of the experience accumulated during the operation of the SNA, several sources of unintentional interference have been identified.

Unintentional artificial interference is caused by the radio transmitters radiation, which can generate signals with an undesirable level of power in the *L*-band. The identified artificial unintentional interference is generated by radio lines, harmonics of television channels, requested proximity navigation signals, harmonics of existing VHF radios, satellite communication system, radar stations of the air traffic control system.

The likelihood of such interference depends on the state regulations in the spectrum and frequency distribution use, as well as ensuring compliance with established rules in each state or region.

Portable electronic devices can cause interference for SNS and other navigation systems as well.

Intentional interference. Due to the low power of the SNS signals, there is a possibility of their suppression by low-power transmitters.

Intentional interference (jamming) is radio interference created by a specially designed source that intended to disrupt the operation of the SNS users' equipment. Intentional interference should also include any actions for disrupting the operation of the SNS, including attacks on satellites and ground control infrastructure.

Another type of intentional interference is radio disinformation that is a method to make the SNS receiver link to false signals similar to standard ones, and slowly descend from an assigned direction so that a sufficiently large time interval has elapsed before an outside interference is detected.

Sauta O.I. et al., *Principles of Radio Navigation for Ground and Ship-Based Aircrafts*,
Springer Aerospace Technology, https://doi.org/10.1007/978-981-13-8293-2_7

Atmospheric interference. There are two ionospheric phenomena that should be taken into account: Rapid and significant changes in the ionosphere state and scintillation. Changes in the ionosphere state lead to errors in the distance measuring, which are taken into account during the system designing.

In the presence of occasional powerful adverse factors as: at geomagnetic perturbations and bursts of radio emission from the Sun, it is possible both a significant deterioration in the location of the aircraft and loss of the SNS signal from one or more satellites.

Various types of augmentation systems are used to obtain a more stable, non-distorted signal from navigation satellites, and to improve accuracy, as follows:

A wide-area Satellite-Based Augmentation System (SBAS);
A Ground Regional Augmentation System (GRAS)
A local Ground-Based Augmentation System (GBAS).

In ground-based and satellite augmentation systems, an important thing is to increase the accuracy of the aircraft position determining by a differential method of correcting the satellite signal (Fig. 7.1).

The essence of the differential method of adjusting the satellite signal is based on the relative constancy of a significant part of the SNS error both in time and space. The implementation of the differential method is possible if there are two receivers, one of which is on the ground and the other is on board the aircraft. The geodetic coordinates of the ground receiver (referred to as the control station) in the selected coordinate system in WGS-84 or PZ-90.11 are known with high accuracy. The accuracy of the monitoring station coordinates determining should not be worse: in latitude and longitude—5 cm; in height relative to the surface of the

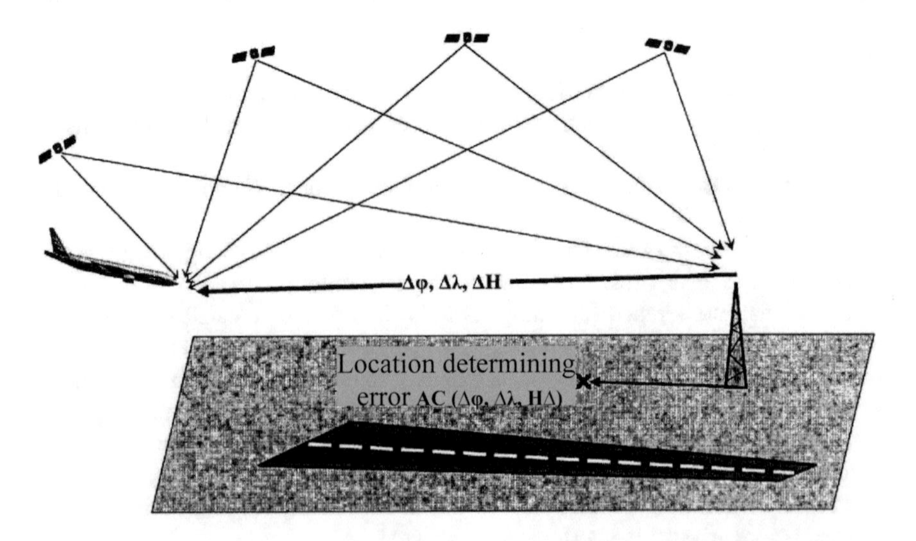

Fig. 7.1 Differential method of accuracy increasing for determination of aircraft location

ellipsoid—0.5 m. The monitoring station receives satellite signals and determines the current coordinates, which are further compared with the coordinates of the reference monitoring station. The corrections to the coordinates for the given area and for the current time are determined by the results of the comparison. The received corrections are transmitted from beacons to consumers via a special radio communication line and, after their processing, differential corrections, information on the system integrity, and other service messages are sent to the SNS onboard receiver. Differential correction signals from radio beacons are transmitted at frequencies of 283.5–325 kHz. Radio signals at these frequencies are subject to reflection from the Earth's surface. Therefore, hilly and mountainous terrain usually does not affect the reception of the signal.

The differential mode is used in both ground and satellite augmentation systems. The use of differential corrections makes it possible to significantly improve the accuracy, and most importantly, the integrity of the aircraft positioning, and thus use the SNS to perform an accurate approach for landing.

In practice, local and global differential GNSS are used.

Local differential GNSS is a variety of differential GNSS in which measured differential corrections can be used to provide appropriate flight stages within a limited geographic area with a radius of 350 km.

Global differential GNSS is a variety of differential GNSS in which calculated differential corrections can be used to provide appropriate flight stages within a limited geographic area with a radius of 350 km.

The most important feature of using SNS is the presence of the computer database of navigational data (NAVDATA) in the onboard equipment. The onboard navigation database includes three interrelated databases:

The main navigation database;
The base of the user's points;
The base of the user's routes.

The main navigation database

The main navigational database covers the Earth territory from latitude 74°N to latitude 60°S. It is possible to use the navigation system outside of the abovementioned territory, but it is necessary to manually input the magnetic declination for correct calculation of magnetic bearings and magnetic track angles.

The volume of navigational database and its content vary slightly depending on the SNS type.

The database is created for regions.

The navigational database includes information on navigation points of the following categories:

- airports;
- VOR beacons;

The general information about airports is as follows:

- identifier;
- name;
- the nearest major city and state;
- latitude and longitude;
- absolute altitude;
- frequencies of communication channels.

Depending on the type of onboard system, the following additional information about airports may be included:

- identification (civil or military);
- information about the runway (traffic volume, elevation, dimensions);
- landing system (lighting and radio equipment);
- presence of a landing radar;
- a zone with a special flight procedures (if the airport is within such a zone);
- difference in time;
- information on airfield services (availability of fuel brands and other services).

Besides, the user can input additional information about each airport consisting of approximately 30 characters into some onboard systems.

The general content of information on radio equipment: VOR, VOR/DME, DME, VORTAC, TACAN, GBAS, etc.:

- identifier;
- GBAS ID;
- positioning channel numbers and GBAS landing channel numbers;
- name;
- VOR or GBAS location country;
- frequency;
- availability of a DME beacon;
- latitude and longitude.

The navigational database, depending on the type of navigation system, may also contain the following information:

- minimum safe altitudes;
- areas controlled by ATS units and airspace classes;
- zones required special flight procedures;
- GBAS landing data units;
- communication means frequencies for contact with ATS.

The main database of navigational data is updated every 28 days, or every 56 days in the United States. The navigational database is supplied on magnetic media: floppy disks or special cartridges inserted into the onboard navigation system unit. The content of the navigation database information can not be changed or edited by the user.

7.2 Principle of Operation in Relative Mode

One of the most important issues for radio navigation is to improve the accuracy of the object or the carrier of onboard RNS equipment location. The radio navigation differential mode, which recently began to apply, makes it possible to significantly increase the accuracy of the object positioning in the absolute coordinate system, but it requires the input of corrections into the object onboard equipment that performs navigation determinations, the organization of communication channels with objects, and the creation and operation of monitoring stations known by high accuracy absolute coordinates.

Unfortunately, it is not always possible to ensure the presence of a monitoring station in the range of acceptable spatial correlation with the onboard equipment of the consumer performing navigation determinations in the phase field of a particular RNS. This may be purely geographical in nature when, for example, onboard systems installed on a helicopter for ice situation reconnaissance, operating with the signals of high-precision marine-used SRS of mid- and low-frequency navigation systems with a range of up to 1000 km, on the one hand, should ensure an accurate investigation of ice conditions and unconditional return of the helicopter to the icebreaker, and on the other hand, the sea area does not allow to install a stationary monitoring station with known absolute coordinates.

It can also be economic reasons, when, for example, each episodically used, poorly equipped airfield cannot be equipped with a stationary monitoring station to provide an accurate approach for landing.

There are also possible reasons of a temporary nature, when for emergency rescue operations with use of mobile objects equipped with onboard equipment of the consumer, there is not enough time for the deployment and accurate geodetic reference to the stationary monitoring station.

It should also be taken into account that when implementing a differential mode in onboard equipment, and especially in the monitoring station equipment, additional computational resources are required to ensure identification, transmission, and recording of correcting information.

At the same time, an analysis of tasks that are to be solved by use of radio navigation systems makes it possible to determine that in many cases there is no need to know the exact absolute coordinates of objects. It is sufficient to ensure mutual coordination of mobile objects or moving objects with respect to a fixed reference point located in the same radio navigation field. This approach has led to the idea of implementing a regime of relative radio navigation for global RNS, as well as for long-range (medium) range RNS.

The essence of the relative radio navigation method is as follows. The onboard consumer equipment coupled with the data transmission equipment is used at the reference point for transmission of coordinates fixed by it. Relative to this reference point, the other moving objects are coordinated, for which they use a new relative coordinate system centered at the reference point. Since the exact coordinates of the reference point position are unknown, it is impossible to realize a differential method.

Thus, the estimate vector of the current relative coordinates of each mobile object is

$$\hat{\delta}(t) = \widehat{X_{\text{п.о}}}(t) - \widehat{X_{\text{о.п}}}(t-\tau) \quad (7.1)$$

where τ is the time difference of navigation definitions on the mobile object and at the reference point; $X_{\text{п.о}}(t)$, $X_{\text{о.п}}(t-\tau)$ are the estimate vectors of the current coordinates of the mobile object and the reference point at instants of time t and $t-\tau$.

It follows from (7.1) that, similar to the differential mode, the error vector of the current relative coordinates of the mobile object is

$$\Delta\delta_{\text{x}}(t) = \Delta\text{X}_{\text{п.о}}(t) - \Delta\text{X}_{\text{о.т}}(t-\tau) \quad (7.2)$$

In case of using a differential distance radio navigation system, the measured value of the ith radio navigation parameter can be represented in the form

$$\psi_{i\text{изм}} = \psi_{ic} + \Delta\psi_{i\text{ин}}(t) + \Delta\psi_{i\text{пом}}(t) + \Delta\psi_{i\text{нест}}(t) \quad (7.3)$$

where ψ_{ic} is a constant component of the measured value; $\Delta\psi_{i\text{ин}}(t), \Delta\psi_{i\text{пом}}(t), \Delta\psi_{i\text{нест}}(t)$ are independent, random, Gaussian stationary time functions: $\Delta\psi_{i\text{ин}}(t)$—a function due to the instrumental error of consumer equipment; $\Delta\psi_{i\text{пом}}(t)$—a function due to interference conditions; $\Delta\psi_{i\text{нест}}(t)$—a function due to the instability of the radio navigation field.

Each random component is characterized by a zero mathematical expectation, corresponding variances, and correlation functions which are the same for each i.

Taking into account the calculated value of the radio propagation velocity V_p and its true value for the signal from the ith station with the weather conditions unchanged—V_{pi} (from the master station—V_{po}) taken for onboard equipment, for a fixed point, we can write

$$\psi_{ic} = 2\pi f\left(\frac{D_i}{V_{pi}} - \frac{D_0}{V_{po}}\right) + \Delta\psi_{iu} \quad (7.4)$$

$$\psi_{i\text{расч}} = 2\pi f\left(\frac{D_i + d_i}{V_{pn}} - \frac{D_0 + d_0}{V_{pn}}\right) + \Delta\psi_{in} \quad (7.5)$$

where $\Delta\psi_{i\text{п}}$, $\Delta\psi_{iи}$—accepted and true base-code delays; D_i, D_0—true distances to ith and leading reference stations; $\psi_{i\text{расч}}$—the estimated value of ith radio navigation parameter; d_i, d_0—errors in knowledge of the coordinate and leading reference stations in the direction of the radio signal propagation.

When using the method of least squares to process measurements, the mathematical expectation of the error vector of the relative coordinates for the difference-ranging three-station RNS will have the form

$$M[\Delta\delta_x] = H_{no}^{-1}\left[L_{ino} + k\delta\psi_{16} - (d_{ino} - d_{o.no})\right] - H_{om}^{-1}\left[L_{iom} + k\delta\psi_{16} - (d_{iom} - d_{o.om})\right] \quad (7.6)$$

where k is the dimension factor; $\delta\psi_{16} = \Delta\psi_{1и} - \Delta\psi_{1п}$—error vector of base-code delays; $\delta\psi_{16}L_i = D_i(\frac{V_{рп}}{V_{pi}} - 1)$—vector of differences between the measured and calculated values of the ith radio navigational parameter; H—matrix of measurement gradients (2 × 2 for three-station difference-distance RNS).

$$\Delta D_i = |D_{iпо} - D_{iот}| \ll D_i, \quad i = 0,1,2 \quad (7.7)$$

we can assume that

$$\mathrm{H}_{\text{по}}^{-1} \approx \mathrm{H}_{\text{от}}^{-1} = \mathrm{H}^{-1} \quad (7.8)$$

$$V_{piпо} \approx V_{piот} = V_{pi} \quad (7.9)$$

In this case

$$M[\Delta\delta_x] = H^{-1}\left[(D_{ino} - D_{iom})\left(\frac{V_{pn}}{V_{pi}} - 1\right) - (D_{o.no} - D_{o.om})\left(\frac{V_{pn}}{V_{po}} - 1\right)\right] - H^{-1}[(d_{ino} - d_{iom})]. \quad (7.10)$$

With the independence of instrumental measurement errors in consumer equipment, the identity of the remaining components and the independence of the $\psi i_{изм}$ vector components, the correlation matrix of the random $\Delta\delta_x$ vector has the form

$$\mathrm{P}(\tau) = 2k^2[H^{-1}H_\tau^{-1}][\sigma_{\text{ин}}^2 + \sigma_{\text{пом}}^2 - R_{\text{пом}}(\tau) + \sigma_{\text{нест}}^2 - R_{\text{нест}}(\tau)] \quad (7.11)$$

where H_τ^{-1}—transposed inverse matrix of measurement gradients.

When $\tau = 0$ expression (7.11) takes the form

$$P(0) = 2P_o(0) \quad (7.12)$$

where $P_o(0) = k^2[H^{-1}H_\tau^{-1}]\sigma_{\text{ин}}^2$—correlation matrix of instrumental errors of the consumer onboard equipment in the absolute coordinate system.

In the absolute coordinate system, for simplicity of estimating the accuracy of position determination, it is customary to use non-correlation matrices, and radial root mean square error including instrumental σ_{M} determined by the expression

$$\sigma_{\text{M}} = [trP_0(0)]^{1/2} \quad (7.13)$$

Then, taking into account (7.12), (7.13) by analogy for the relative coordinate system

$$\sigma_{\text{OTH}} = \sqrt{2}[trP_0(0)]^{1/2} = 1.4\sigma_{\text{M}} \quad (7.14)$$

Thus, the radial root mean square error of estimating the relative coordinates at $\Delta D \rightarrow 0$ and $\tau \rightarrow 0$ is determined mainly by the instrumental error of the consumer onboard equipment (7.14). This is significantly lower than the analogous estimates for absolute coordinates due to the compensation of systematic and highly correlated random errors included in them.

The relative radio navigation mode can be used in the following cases:

At the group movement of ships, vehicles, airplanes, and other air means to ensure that they do not have large safe distances between them while moving;
In systems of preventing collisions of air or sea objects;
When solving the tasks of ships, aircraft, and land transport proximity;
When organizing with the help of the latest rescue and recovery works, etc.

Such wide possibilities of relative radio navigation application are explained by high accuracy characteristics of the regime. Indeed, the theoretical results obtained have got repeated experimental confirmation in the derivation in various ways: Route or permanent navigation mode of mobile objects to points with coordinates previously determined by similar consumer onboard equipment.

For example, as practice shows, for long-wave RNC having in real conditions for absolute coordinates the average error value of about 150 m with an instrumental mean square error of about 20 m, the error of inference lies in the range from 10 to 60 m for a region with a geometric factor of about 1.3 and distance from the reference stations 500–700 km.

For practical implementation of relative determinations by signals GPS RNS, mainly in the United States, a number of specialized sets of equipment were developed: SERIES, NGPS, V-1000, MACROMETER, GEOSTART, etc.

Features of this equipment are the ability to record measurements performed at remote locations, with their further joint processing; carrying out of measurements on time intervals from several hours to several days; wide use of the phase as a measured parameter, using a variety of methods to eliminate the phase divergence and frequencies of the navigation satellite generators and the points to be determined, as well as to disclose the ambiguity of phase measurements, etc. It is also noted that the most common relative determinations were obtained in solving problems of geodesy and geodynamics.

References

1. Yarlykov MS (1995) Statistical theory of radio navigation. M. Radio and Communication, 344 pp
2. Shatrakov YG (1999) Estimation of location error in the PI of the difference-ranging system and their dependence on the hyperbolic position lines. Col. Ed. Issue 1. VIMI, C 58–95
3. Shatrakov YG (1990) Radio navigation systems of mobile objects. M. MRP, 92 pp
4. Shebshaevich VS et al (1989) Differential mode of network satellite RNS. Foreign radio electronics, No. 1, C 5–32
5. Shebshaevich VS (1992) Network satellite RNS. M. Radio and Communication, 280 pp
6. Shatrakov YG et al (1983) Correlation errors in VHF goniometric radio engineering systems. M. Sov. Radio, 280 pp
7. Sauta OI et al, under editorship of Shatrakov YG (2016) Development of navigation technologies for improving flight safety. SPb, GUAP, 299 pp
8. Zavalishin OI, Korchagin VA, Lukoyanov VA, Mironov MA (2003) Monitoring of signal quality at control-correcting stations, providing differential operation of consumers of satellite radio navigation systems. Radiotekhnika 1:35–44
9. ICAO annex 10 volume 1 aeronautical telecommunications—radio navigation aids

Chapter 8
Augmentation Systems to Ground-Based GNSS

8.1 Principles of Creation of Satellite Augmentation Systems

Since the deployment of the global navigation satellite systems in the early 1990s (GPS in the US and GLONASS in the USSR), the ideology of their use in aircraft has evolved from a radical replacement of all existing navigation aids to the concept of sharing diverse navigation sources. At the end of the first decade, twenty-first century, after accumulating rich experience in the practical operation of the GPS system and the orbital grouping of GLONASS was restored, each of these systems is considered already as the single one, but as the main or additional navigation tool, requiring the use of duplicate navigation systems [1–9].

In international practice, a new concept has emerged—The Global Navigation Satellite System (GNSS), which includes two systems: GPS and GLONASS.

In accordance with the new concept, GNSS is considered as the main navigational aid for flights over the ocean and as an additional system for other applications, and in the presence of functional ground-based add-ons, as the main means for all stages of the flight.

To solve the specific tasks of navigation with GNSS, the International Civil Aviation Organization (ICAO) has developed a concept for GNSS ground-based (GBAS) and space-based (SBAS) functional additions that allow GNSS to become the primary navigation tool for airport areas, with non-precision and precision approach, with ground navigation at the airport, as well as for air traffic control using automatic dependent surveillance (ADS-B).

Sauta O.I. et al., *Principles of Radio Navigation for Ground and Ship-Based Aircrafts*,
Springer Aerospace Technology, https://doi.org/10.1007/978-981-13-8293-2_8

This section will consider in detail the augmentation system of ground-based GNSS (GBAS in ICAO terminology) with a zone of action that provides a solution to the navigation task in the airfield area and the landing of aircraft in accordance with the requirements for category I[1] landing systems. To refer to the terrestrial part of GBAS, we will use the term LAAS (Local Area Augmentation System). The onboard part of GBAS will be denoted as GNSS/LAAS (GLS).

In accordance with the plans of the US Federal Aviation Administration (FAA), the task of ensuring the landing of aircraft in the category I of complexity in the United States was to be solved in 2001, and by 2005, it is planned to provide landing in the complexity Categories II and III. Work in this way has been going on for many years in the United States. In [7], for example, the results of tests of the landing system performed by 1997 are described. With regard to the provision of Category I landing, the task in the USA and a number of European countries has been fulfilled. However, landing on the complexity Categories II and III has not been practically implemented so far due to the complexity of providing equipment certification procedures and the lack of harmonized regulatory documents. Standards for Categories II and III landing were approved in 2018.

The practical implementation of Category I satellite systems for landing in Russia began in 2006, after certification and installation of the LKKS-A-2000 manufactured by the company NPPF Spectrum in several airports and the completion of certification procedures for the onboard equipment of GNSS/LAAS. By early 2018, about 120 Russian airports were equipped with LKKS-A-2000, and serial production of onboard GNSS/LAAS equipment began [9].

8.2 General Information on GBAS

The general ideology of GBAS is based on the concept of differential subsystems and consists of the following: At the location of several GNSS receiving antennas (used by the LAAS linear measuring instruments), the coordinates of which in the geodetic coordinate system are determined with high accuracy, GNSS signals are received and processed, and the parameters of integrity and corrective information are formed (usually corrections to the measured pseudo-ranges and their rate of change). Then, this corrective information is transmitted via the available ground-to-aircraft communication channel to the onboard GNSS receiver, where it is used to exclude strongly correlated pseudo-range measurement errors from the measurements. As a result, the consumer onboard the aircraft receives adjusted coordinates and time. It is obvious that the effectiveness of the differential method application depends

[1] As defined by ICAO, the Ccategory I landing system provides guidance to the aircraft from the boundary of its coverage area to the point at which the course line crosses the glide path at a height of 60 m above the horizontal plane passing through the threshold of the runway. This definition does not exclude the use of category I systems below a height of 60 m in the presence of a visual orientation. For a cCategory I landing system using GBAS, the error in determining the horizontal coordinates should be 16 m (95%), and the altitude error should be 4–6 m (95%).

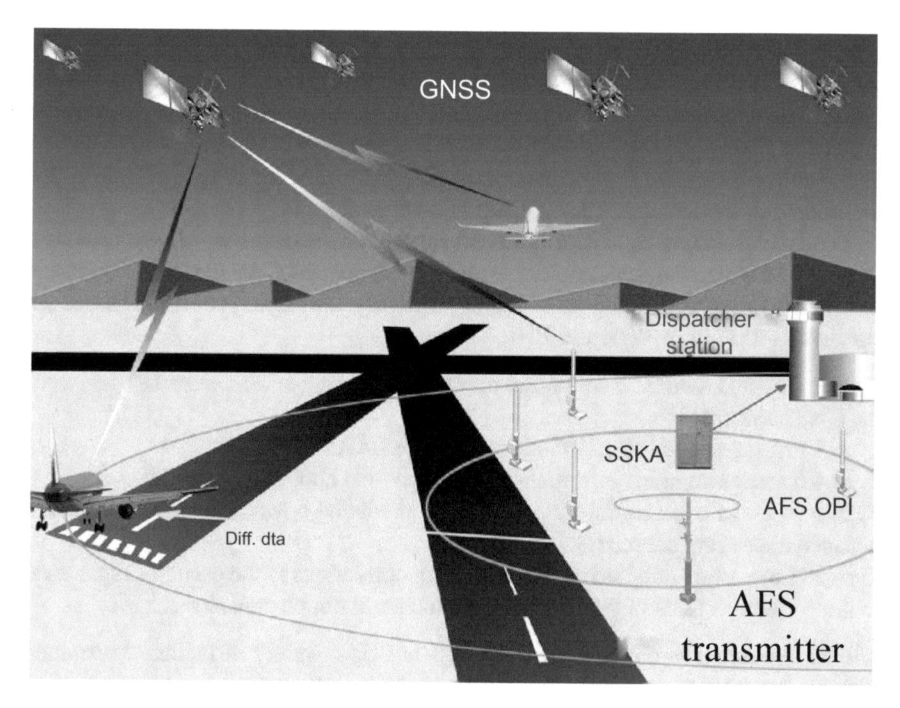

Fig. 8.1 Satellite landing system with use of GBAS

on the degree of spatial and temporal correlation of the errors on the LAAS and on onboard the aircraft. With a strong correlation, the systematic part of the error will be eliminated almost completely; with a weak correlation, a residual error will appear.

To meet the requirements for ICAO Category I landing systems, regulatory requirements for the GBAS system have been developed, which will be discussed below.

For a clearer general view of the satellite landing system arrangement, let us consider the illustration in Fig. 8.1.

Figure 8.1 shows the LAAS station located not far from the runways (RSs) of the airfield, and the antenna-feeder system of reference receivers (AFS LAAS) that are part of the LAAS. The information from all reference receivers that receive navigation signals from GNSS navigation satellite devices (NS) is processed on the LAAS and then transmitted to the LAAS transmitter and transmitted to the consumers of the corrective information via the antenna-feeder device of the transmitter (AFS transmitter). Corrective information received onboard the aircraft enters the onboard equipment of GNSS/LAAS, where it is used to specify the coordinates and speed of the aircraft.

As GBAS is designed to provide positioning services in the airfield vicinity and for ICAO Category I landing, and later for higher categories, in performing the differential data transmission "earth-to-aircraft", much attention is paid to interference protection and interference resistance of this channel. We will return to this issue later, and now we will briefly consider the composition of the messages transmitted to the aircraft by the LAAS through the radio channel.

The LAAS should be able to transmit up to 256 messages of various types via the radio channel. Such possibility is put on the prospect, including for securely landing by Categories II and III of ICAO. Currently, only five types of messages are standardized: 1, 11, 2, 4, 5, and 101.

Type 1 message there is information on the pseudo-range corrections for the NS of the GNSS system.
Type 2 message contains information about the LAAS itself.
Type 4 message contains information about the final approach segment.
Type 5 message contains information on the predicted operational availability of a distance measuring source (GNSS NS).
Type 101 message contains information about corrections to the pseudo-ranges used in the regional extension system GRAS, which will not be considered here.

The configuration of a concrete LAAS depends on a variety of factors determined both by the characteristics of the location of the LAAS, and the predicted mode of the LAAS implementation. But in any case, the module of reference receivers and the transmitter of the "radio channel earth-board"—transmitter VDB will be present in the LAAS structure.

The generalized LAAS structure is shown in Fig. 8.2.

The structure of the onboard subsystem GBAS is shown in Fig. 8.3.

In general, the structure of the GNSS/LAAS equipment depends on the structure of the aircraft onboard complex. For example, the antenna of the VDB receiver can be used as a beacon antenna of the ILS type instrumentation landing system, and the FMS control panel can be used as an operating and display unit.

The main functions of onboard GNSS/LAAS equipment are [1]: Reception of GNSS signals, reception and processing of LAAS messages, selection of positioning services (RSDS) for RNAV GNSS, RNP APCH, and approach (RPAS), formation of parameters for accurate guidance (ILS-like signals), detection of precise landing approach area (PAR), formation of navigation parameters (coordinates, speeds, and time) and alarms, as well as indicators of accuracy and integrity calculated using standardized algorithms.

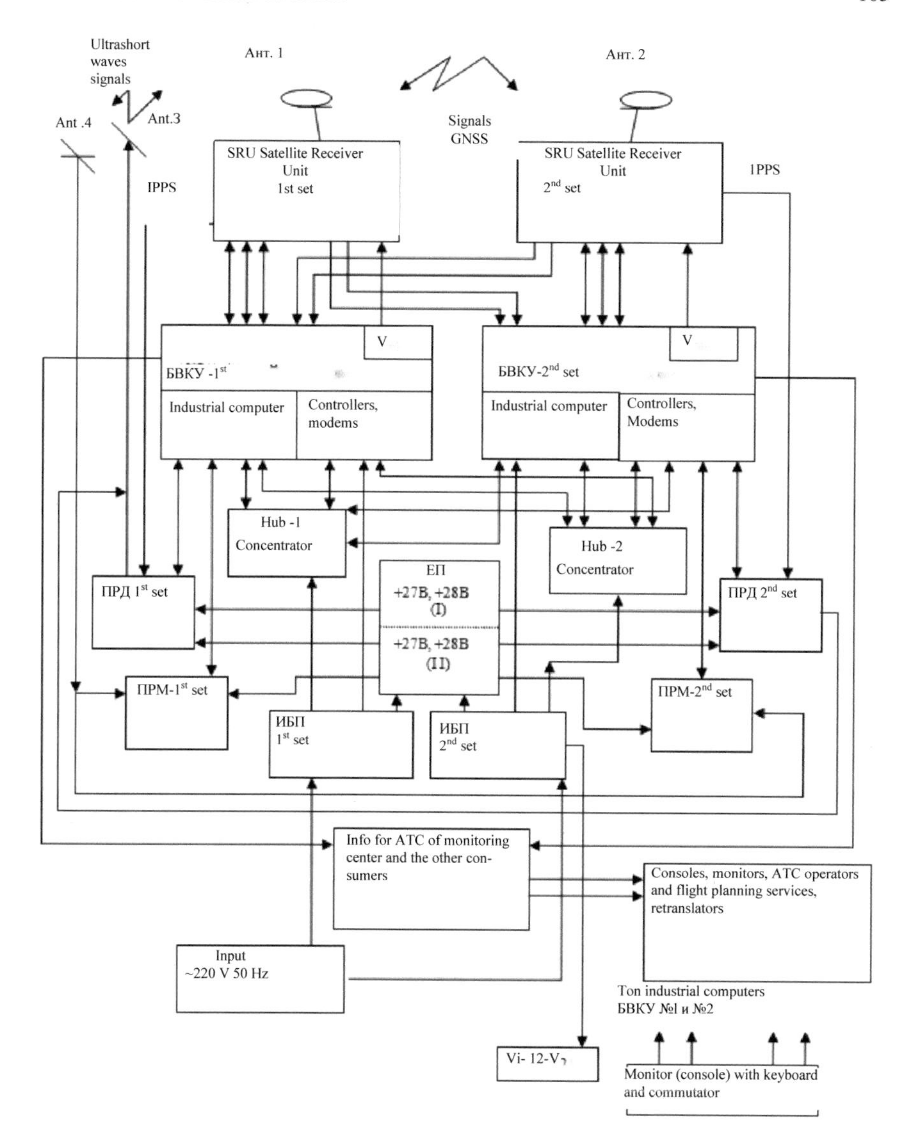

Fig. 8.2 Generalized LAAS structure

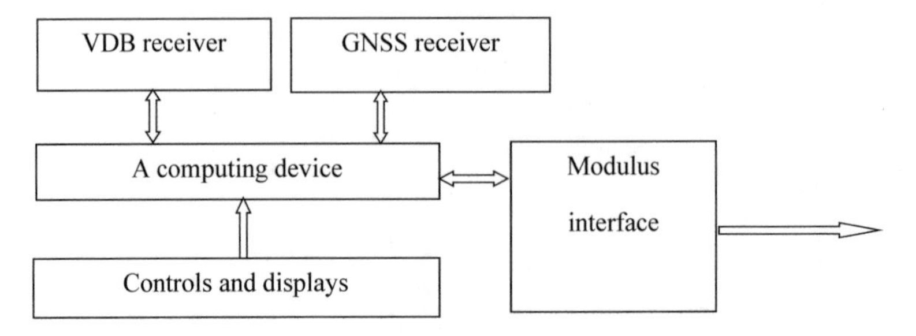

Fig. 8.3 Onboard equipment GBAS

8.3 Purpose, Basic Requirements and Functions of GBAS

Let us briefly consider the characteristics of both subsystems of GBAS: ground-based LAAS and onboard GNSS/LAAS.

In accordance with ICAO Regulations [1], "…The GBAS system is designed to support all types of approach, landings, departures and ground operations and can provide operations on the route and in the airfield area" and must perform the function of supporting precise approaches and categorized of landings of aircraft to aerodromes and sites, as well as supporting other vehicles and procedures that require precise location, provided they are within the coverage area of the LAAS.

The functional capabilities and characteristics of the LAAS for the stages of the route, the flight in the airfield area, the approach and landing procedure depend on the category of LAAS (A, B, C, or D). Category A characterizes LAAS ensuring compliance with the minimum requirements for precision and reliability characteristics; Category D characterizes LAAS satisfying the maximum requirements including ensuring landing by Categories II and III of ICAO.

The functionality and characteristics of the onboard GNSS/LAAS equipment depend on its class (A or B). Class A characterizes onboard equipment with minimum requirements for standard errors of corrective pseudo-range determination, and Class B characterizes high-precision equipment designed to support ICAO Categories II and III landings.

The main functions of the LAAS are

- Generation of messages of types 1, 2, 4, 5;
- Transmitting messages over a radio channel;
- Integrity control: of data of observed NS, differential data, radio channel, and messages transmitted thereon;
- Control of continuity of generated and transmitted data;
- Control of own performance;
- Control of the equipment operation and modes of operation of LAAS;
- Registration of messages, parameters, and operating modes of the LAAS, failures, defects, and other operational disturbances, external control actions, and environmental conditions transmitted via the radio channel;

- Reception, storage, updating, and transfer of auxiliary, service, and other information via the LAAS to an airport traffic control tower.

The main functions of onboard GNSS/LAAS equipment are [1]

- Guidance in the implementation of terminal procedures and approach;
- Issuance of data on deviations from the landing path;
- Output of position, velocity, and time (PVT) data both in the presence of differential corrections and in the absence of them.

8.4 Technical Requirements and Characteristics of LAAS

As mentioned earlier, the ground-based augmentation system (GBAS) consists of ground and airborne elements. The ground subsystem of the GBAS is LAAS that typically includes one active ultrashort wave transmitter (VDB transmitter), a transmitting antenna, and several GNSS reference receivers. The LAAS may include multiple VDB transmitters and antennas that use one common identifier (GBAS ID) and a carrier frequency, and also transmit identical data. One LAAS can support all airborne subsystems within its coverage area, providing the aircraft with data for approach, corrections and integrity information for visible NSs.

LAAS provides two types of services: Approach and location. When approaching, guidance is provided on the final precision approach sections, as well as when performing a non-precise approach (NPA) or an uncategorized approach with vertical guidance (APV) and within the LAAS operational area. It should be noted that a Type 4 message is transmitted by the LAAS in the event that precise approaches are made in its coverage area (Categories I, II, III).

The ground and airborne subsystems of the GBAS provide position information in the horizontal plane in order to provide zone navigation (RNAV) operations within the coverage area when determining a location. The principal difference between the precision approach functions and RNAV is the different operational requirements associated with the specific operations provided, including different data integrity requirements.

One of the most important functions of the LAAS is the transmission of additional parameters of the location limit errors due to de-correlation of pseudo-range errors for the diversity of the receiver devices due to errors in the ephemerides. This function is necessary for locating, but is optional when providing approaches. If these additional parameters of location error limit are not transmitted, the LAAS is responsible for ensuring the integrity of the ephemeris data of the ranging sources, not relying on airborne calculations and applying the ephemeris limit values.

There are several possible configurations of LAAS that meet international standards [1]. Among them are the following:

(a) Configuration that provides only a precise approach for Category I landing.
(b) Configuration that provides a precise approach for landing in Category I and APV, and also transmits additional parameters of the location limit errors due to errors in the ephemerides.
(c) Configuration that provides a precise approach in Category I and APV landing and positioning, while transmitting the parameters of the location limit errors due to errors in the ephemerides.

The LAAS performs data transmission by using a radio signal with horizontal or elliptical polarization (GBAS/H or GBAS/E). This allows the transmission to be adapted to the operational requirements of the user community, i.e., to receive data onboard the aircraft which are equipped with antenna-feeder systems of various types.

Currently, it is assumed that most aircraft will be equipped with horizontal polarization receiving VDB antennas, which can be used to receive the VDB signal from both GBAS/H and GBAS/E equipment. At the same time, aircraft equipped with vertical polarization antennas (due to restrictions on location or economic reasons) will not be able to use data from the LAAS with GBAS/H equipment. Aircraft operators who use receiving antennas with vertical polarization need to take this information into account when flying.

LAAS (transmitter VDB) provides emission of data and corrections to the range-finding signals of GNSS by transmitting digital data in the radio frequency range of 108–118 MHz. The lowest assigned frequency is 108.025 MHz, and the highest assigned frequency is 117.950 MHz. The separation between the assigned frequencies (channel separation) is 25 kHz.

Considering the high requirements for data integrity characteristics in the implementation of precise approaches, the LAAS transmitter is subject to rather stringent technical requirements. For example, carrier frequency stability is maintained in the range of ±0.0002% of the assigned frequency; data coding is performed by means of a bit-wise phase shift; the messages themselves are formed as symbols, each of which consists of three consecutive bits of the message, and the symbols are transformed into a differential 8-bit level format (D8PSK) by means of a phase shift of the carrier frequency of 45°.

The magnitude of the error vector of the transmitted signal does not exceed 6.5% of the RMS value (1σ). The symbol transmission rate is maintained at 10,500 symbols/s (with an accuracy of ±0.005%) and provides a nominal data transfer rate of 31,500 bps.

The LAAC transmitter uses a time-division multiple access (TDMA) system, which is based on frames and time slots. Duration of each frame is 500 ms. In each one-second UTC era, there are two such frames. The first of these frames begins at the beginning of the UTC era, and the second begins is 0.5 s after the start of the UTC era. The frame is time multiplexed so that it consists of eight separate time intervals (A–H) of 62.5 ms duration. In each set time interval, max. one packet is contained.

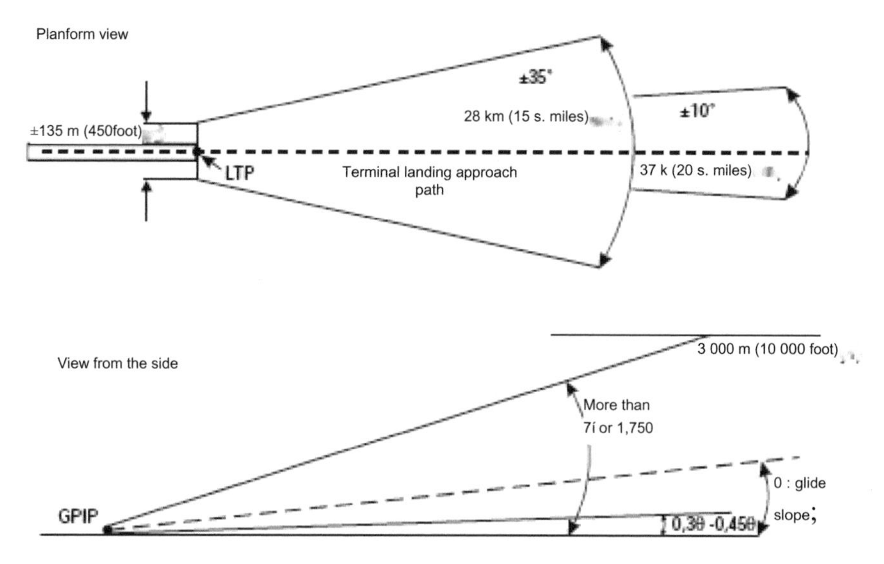

Fig. 8.4 The minimum coverage area of the LAAS

To initiate the use of a slot (s), the LAAS transmits a packet in a given time slot in each of five consecutive frames. The ground system transmits the packet in at least one of every five consecutive frames in each used time slot.

The data transmitted by the LAAS are encoded and simultaneously protected using a fixed (255, 249) Reed–Solomon code of fixed length with a defining polynomial of the 8th degree [1]. When encoding, data is grouped into 8-bit symbols of the Reed–Solomon code. This code can correct up to three symbol errors inclusive.

Moreover, all blocks of messages transmitted by LAAC include a 32-bit check cyclic redundant code (CRC), which provides the probability of missing an error of max. 2.3×10^{-10}.

The minimum coverage of the LAAS for landing approaches is shown in Fig. 8.4.

In the case when additional parameters of the limiting position errors are transmitted by LAAS (see above), the differential corrections can be used only within the maximum used distance (D_{max}) defined in the type 2 message.

The coverage area of the LAAS required to ensure the location of the aircraft using GBAS depends on the planned specific operations. The optimum coverage for this service should be omnidirectional in order to ensure that the aircraft position is outside the coverage area of the precision approach.

The use of location information is limited to the maximum distance used (D_{max}) that determines the range within which the required integrity is ensured and differential corrections can be used, either for locating or for precise approach. In this connection, location-based operations can be predicted only in the coverage area within the D_{max} range.

Since the desired area of location determination using GBAS may be greater than the coverage of one LAAS, an LAAS network may be used to provide it. These stations can transmit on one frequency and use different time intervals (which are currently defined to be eight) at neighboring stations to avoid interference, or they can transmit data at different frequencies.

To meet the specified landing requirements for ICAO Category I, the VDB receiver, from the GNSS/LAAS equipment, should also meet fairly high requirements. Let us briefly consider these requirements.

To receive data from the LAAS, the frequency range of the onboard VDB receiver should be 108,000–117,975 MHz in 25 kHz increments, and search and capture of LAAS signals should be ensured when the frequency is shifted within the band ± 418 Hz from the nominal set frequency due to the aircraft movement and instability of the frequencies of the LAAS transmitter.

The onboard VDB receiver should be set to the frequencies of the LAAS transmitter, depending on what is received via the control channel. In this case, the output of messages received from the LAAS at a new frequency should start within 3 s from the moment the command is received to change to this frequency.

The intensity of incorrect message reception in the onboard VDB receiver should not exceed one incorrectly received message per 1000 full-length messages (222 bytes).

It is an extremely important task for the onboard VDB receiver to provide interference resistance. It should provide interference suppression on the working channel in the presence of signals from systems such as VOR, ILS, and VHF FM broadcast signals.

To ensure all types of aircraft traffic paths in the airfield area, the directional pattern of the AFS of the onboard VDB receiver in the horizontal plane should be omnidirectional.

8.5 The Concept of System Data Integrity

For precise approach and operations based on positioning using GBAS, different levels of integrity are defined.

Integrity in aviation systems is understood as a measure of trust, which can be attributed to the correctness of information issued by the system as a whole. Integrity includes the system ability to provide the user with timely and valid warnings (alarms) [1].

The risk of loss of signal integrity in ICAO Category I landing space is 2×10^{-7} for one approach, the duration of which usually does not exceed 15 min. This value can be visualized as follows: When executing 10 million approaches, it is allowed not to give a warning to the pilot more than two times that the parameters of the navigation system are out of tolerance.

LAASs that are designed to provide other operations based on location should also meet the requirements for the risk of loss of signals integrity in the space established for operations in the terminal area, which is 1×10^{-7} 1/h. In this case, additional measures are required to ensure more stringent positioning requirements.

The risk of signal integrity loss in space is distributed between the risk of integrity loss of the ground subsystem (LAAS) and the risk of integrity loss of the protection level calculated by the onboard GNSS/LAAS equipment. The share related to the risk of the LAAS integrity loss is determined by the failures of the ground subsystem, as well as by failures of the general orbital system of the NS, for example, as a result of a decrease in signal quality or errors in ephemerides. The share related to the risk of the protection level integrity loss is determined by the rare risk of failure-free operation and the case of failure in one of the measurements of the LAAS reference receiver. In both cases, in the onboard GNSS/LAAS equipment, in calculating the level of protection, consideration is given to the effect of the geometry of the spatial location of the NS which signals are used by the aircraft receiver.

The LAAS determines the estimation error for the pseudo-range differential correction (σ_{pr_gnd}) for the GBAS reference point and the additional errors due to difference in the propagation paths of the signals from the NS to the LAAS and to the aircraft in the troposphere (σ_{tropo}) and the (σ_{iono}) ionosphere (so-called spatial de-correlation).

The measurement errors described above for each NS are used in the onboard GNSS receiver to calculate the total error in solving the navigation task and to calculate the protective levels. It is assumed that the errors have a normal distribution.

The lateral protection level (LPL) specifies the allowable error value for a location in the horizontal plane with a probability determined from the above integrity requirements. Similarly, the vertical protection level (VPL) determines the permissible vertical position error value. When performing ICAO and APV Category I precision approach procedures, if the calculated LPL exceeds the specified lateral alarm tolerance threshold (LAL), or if the calculated VPL exceeds the vertical alarm threshold (VAL), a system integrity warning is generated when implementation of a specific operation, i.e., the alarm thresholds are determined based on the operation being performed.

It should be noted that the calculation of the protection levels should take into account the extreme position errors due to errors in the ephemerides.

In general, two levels of protection are defined onboard the aircraft—for the error-free operation of all LAAS reference receivers (hypothesis H0—normal measurements conditions) and for a state where one of the reference receivers has faulty measurements (hypothesis H1—measurement conditions with errors).

ICAO and APV Category I precision approach procedures determine the lateral error bound (LEB) and the vertical error bound (VEB). To determine the location, the horizontal error bound due to errors in the ephemeris (HEB) is also determined.

8.6 Characteristics of Coordinates Determination Accuracy

Determination of the contribution of LAAS to the corrected pseudo-range error (σ_{pr_gnd}) is a rather complex task and is not regulated in the current standards [1]. The task of the LAAS developer is to prove that the value of σ_{pr_gnd} formed on the LAAS and transferred to consumers corresponds to the requirements of the LAAS of the appropriate type (A, B, or C). The sources contributing to this error include receiver noise, multipath, and calibration errors of the antenna phase centers. Receiver noise usually has a normal distribution with zero mean, whereas multipath and calibration of antenna phase centers can lead to a small error in the mean value.

Let us briefly consider other types of errors calculated on the LAAS and transmitted to the aircraft through the radio channel.

First, these are residual ionospheric errors. The ionospheric parameter characterizing this type of error is transmitted in LAAS type 2 messages for simulating ionospheric effects between the LAAS reference point and the aircraft. This error is described fairly well by a normal distribution with zero mean.

Second, these are residual tropospheric errors. Tropospheric parameters are transmitted in LAAS type 2 messages to compensate for tropospheric effects onboard when the altitude of the aircraft and the height of the LAAS reference point are different. This error is also described by a normal distribution with zero mean.

Third, the onboard receiver's contribution to the corrected pseudo-range error. The maximum value of this contribution can be estimated on the assumption that $\sigma_{receiver}$ is equal to RMS_{pr_air} for onboard equipment (GNSS/LAAS) with an accuracy rating of A. For more precise equipment (class B), a more careful calculation is required which is not currently standardized.

Fourth, these are multipath errors due to the aircraft body influence. The contribution to the error of the multipath due to reflection from the aircraft body is determined by the standard equation [1]. Multipath errors due to reflection from other objects are usually not taken into account. However, if the experience of using LAAS at a particular airfield shows that these errors cannot be neglected, they are taken into account by increasing the values of the parameters transmitted by the LAAS, for example, σ_{pr_gnd}. Such conditions usually arise when the airfields are located in conditions of a complex (mountain) terrain or at the presence of a significant number of artificial structures in the vicinity.

Finally, fifth, errors in ephemeris make a significant contribution to the positioning errors. Pseudo-range errors due to errors in the ephemeris (defined as discrepancies in the satellite's true position and satellite positioning based on the transmitted data) are partially de-correlated and therefore will be different for receivers in different locations. In the case when users are relatively close to the LAAS reference point, the residual error due to errors in the ephemerides can be neglected when correcting the measurements and calculating the protection levels. If the aircraft is significantly far from the LAAS reference point, errors in the ephemerides can be accounted for in two different ways:

(a) If the LAAS does not transmit additional parameters of the location error limits in the ephemerides, the LAAS ensures integrity with allowance for errors in the ephemeris of the satellite, without relying on onboard calculations and applying limit errors. This can lead to a limitation of the maximum allowable distance between the LAAS reference point and the aircraft;

(b) If the LAAS transmits additional parameters of the location error limits due to errors in the ephemeris, which allows the onboard receiver to calculate the marginal errors. These parameters are: The coefficients used in the equations for calculating the error limits (Kmd_e_ (), where index () stands for either GPS, GLONASS, POS, GPS, or POS, GLONASS, the maximum distance used for differential corrections (D_{max}) and the parameters of de-correlation of ephemerides (P). The parameter of de-correlation of ephemerides P in a type 1 message characterizes the residual error as a function of the distance between the LAAS reference point and the aircraft. The P value is expressed in m/m. The P values are determined by the LAAS for each satellite. One of the main factors influencing the P value is the LAAS monitoring device. The quality of the ground monitoring device will be determined by the smallest error in the ephemeris (or the Minimum Detectable Error—MDE) that it can detect. The relationship between the P parameter and the MDE for a particular satellite can be approximated using the equation $Pi = MDE_i/R_i$, where R_i is the smallest of the predicted distances from the LAAS antenna (s) reference receiver during the action period of P_i. Depending on the geometry of the satellites, the P parameter values vary slightly. However, P is not required to be dynamically varied by the LAAS. Constant parameters P can be transmitted if they properly ensure integrity. In this case, the readiness indicator will deteriorate slightly. In general, since the MDE becomes smaller, the overall availability of GBAS improves.

8.7 Monitoring Errors in Ephemeris and GNSS Failures

There are a number of monitoring methods to detect errors in ephemeris and GNSS failures. Among these are

(a) An increase in the spacing of the antennas of the LAAS reference receivers. In this case, it is required to use receivers spaced over long distances to detect errors in ephemerides that are not detected by a single receiver. More spacing also improves the detection of the minimum detectable error—MDE.
(b) The use of data from the space-based GNSS augmentation system (SBAS). Since SBAS provides monitoring of satellite characteristics, including ephemeris data, integrity information transmitted by SBAS can be used as an indication of the ephemeris adequacy. The ground-based SBAS subsystem uses GNSS receivers installed with a large territorial separation and, therefore, optimal characteristics of ephemeris control are provided. As a result, it becomes possible to identify small MDE values.
(c) Monitoring of ephemeris data. This method involves comparing the ephemerides transmitted by the NS during the successive passage of satellites over the LAAS. In this case, it is assumed that the only reason for the failure is an error of the ephemeris transmitted by the network of ground control stations and the GNSS information. In order to ensure this method to provide the required integrity, it is

necessary to exclude the possibility of failures due to unauthorized maneuvers of satellites.

Now more attention is paid to monitoring systems. The characteristics of the monitoring device (for example, the MDE detected by it) should be based on the requirements for the risk of integrity loss and the failure model, which protection should be provided by this monitoring device. The error frequency limit for GPS ephemeris information can be determined based on the reliability requirements defined in [1]:

(a) The frequency of main service failures for the orbital constellation as a whole—-max. three per year (global average);
(b) Reliability—min. 99.94% (global average);
(c) Reliability—min. 99.79% (mean for a single point).

The GLONASS monitoring segment that monitors parameters of the ephemeris and the system time, introduces a new navigation message in the event of any unplanned situation. It should be noted that in the event of failures in ephemeris information and time, errors in determining the range are not more than 70 m. The frequency of GLONASS satellite failures, including failures in ephemeris and time information, does not exceed 4×10^{-5} 1/h for any NS.

Since the error in the ephemeris can lead to serious problems in the navigation support, the above GLONASS and GPS characteristics are necessarily taken into account in the integrity control algorithms.

A typical LAAS processes measurements from 2 to 4 reference receivers installed in the immediate vicinity of the reference point. The onboard receiver is protected from large errors or malfunctions in one of the LAAS reference receivers by calculating and applying the integrity parameters B_i transmitted by the LAAS in a type 1 message.

8.8 Continuity of Service

The Continuity/Integrity Index of the ground-based subsystem (GCID) provides a classification of LAASs. If the GCID is set to 1, then the LAAS meets the ICAO Category I or APV precise approach requirements. GCID values 2, 3, and 4 are provided to support future operations with more stringent requirements than those for ICAO Category I operations. The GCID indication is intended to indicate the status of the LAAS that will be used during the approach. The GCID does not provide any indication of the ground-based subsystem capability to provide GBAS positioning services.

LAASs ensuring the implementation of ICAO and APV precise approach for ICAO Category I landing shall [1] ensure continuity of service $1–8.0 \times 10^{-6}$ or more in any 15 s.

LAASs that are designed to support other operations based on location should provide the minimum continuity required for operations in the airfield area, which is $1–1.0 \times 10^{-4}$ for an hour. When continuity ($1–3.3 \times 10^{-6}$ for 15 s) required for an accurate approach by Category I or APV landing is converted to an hourly value, it does not meet the minimum continuity requirement in $1–1.0 \times 10^{-4}$ for an hour. Therefore, additional measures are needed to satisfy the continuity requirement for other operations. One method of ensuring this requirement assumes that there are backup facilities onboard the aircraft, for example, an onboard augmentation system (ABAS) based on inertial systems, and that ABAS provides sufficient accuracy for a particular operation.

8.9 Selecting the Channel and Identifying the LAAS

Due to the fact that the concept of the satellite landing systems based on GBAS introduction includes the requirement of equivalence for the pilot of the aircraft a form of presentation of landing information from any landing system, the channel numbers are used to simplify the interface between the onboard equipment and the pilot, which ensures compatibility with interfaces for ILS landing systems and MLS. The basis for the integration of the aircraft cockpit and the crew interface for GBAS landing is the introduction of a five-digit channel number. Alternatively, an interface based on approach selection using a flight management system (FMS), similar to the existing practice when using ILS, is possible. The GBAS channel number can be stored in the onboard navigation database as part of the named landing approach. The landing approach can be selected by name, and the channel number can be automatically communicated to the equipment, which must select the appropriate data for the approach to the GBAS approach from the transmitted information. Similarly, the use of the location determination function GBAS can be based on selecting a five-digit channel number. This can also simplify the implementation of non-approach operations. In order to facilitate the adjustment of the frequency of onboard GNSS/LAAS equipment, GBAS channel numbers for GBAS ground-based subsystems providing position determination can be included in the additional data block of type 2 LAAS message.

The LAAS channel number in the range of 20,001–39,999 is assigned if the FAS data is transmitted in a type 4 message. The channel number in the range of 40,000–99,999 is assigned if the FAS data associated with the APV is obtained from the onboard database.

The channel number allows the onboard GNSS/LAAS equipment to be set to the correct frequency and select the final approach segment (FAS) data block that determines the desired approach. The required FAS data block is selected by the reference path data selector (RPDS), which is included in the LAAS type 4 message as part of the FAS data definition. A similar scheme is used when choosing a positioning service using the Reference Station Data Selector (RSDS). RSDS is transmitted in an LAAS type 2 message and allows you to select a single LAAS that provides location determination. For LAASs that do not provide positioning and do not transmit additional ephemeris data, RSDS is encoded as 255. All RPDS and RSDS transmitted at a certain LAAS frequency are unique within their coverage area. The RSDS value should not match any RPDS values transmitted.

The GBAS identification is used to uniquely identify the LAAS transmitting messages at a given frequency within the LAAS coverage area. When flying, the aircraft will use the data transmitted by one or more LAASs having a common identifier.

8.10 Final Approach Path

In the final approach stage, the satellite landing system uses a path that is a line in space defined by a point on the runway threshold (LTP/FTP), the flight path alignment point (FPAP), the relative threshold crossing height (TCH), and glide path angle (GPA). These parameters are determined from the information contained in the FAS data block in the LAAS type 4 message or in the onboard aircraft database.

The relationship between these parameters and the FAS path is shown in Fig. 8.5.

FAS data blocks can be stored in a common onboard database and used in the onboard GNSS/LAAS equipment when a Type 4 message is not transmitted.

The LTP/FTP point is usually located at or near the runway threshold. However, due to operational needs or physical limitations, the LTP/FTP point can also be spaced from the runway threshold.

The FPAP point is used in conjunction with LTP/FTP to determine the reference plane for determining lateral departures during the approach. The FPAP point is usually located on the extension of the runway axis beyond the far threshold of the runway (the threshold of the runway opposite the landing runway).

The "Δ-distance offset" parameter specifies the distance from the long runway threshold to the end of the runway to the FPAP point. This parameter is input to

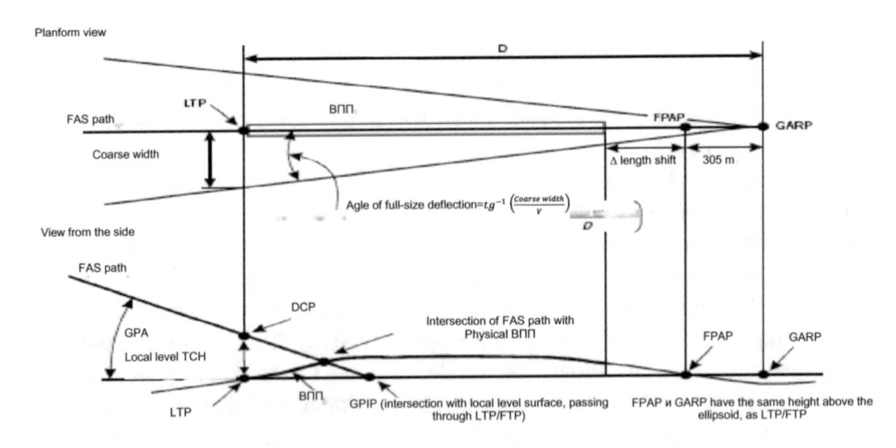

Fig. 8.5 Determining the FAS path

allow the onboard equipment to calculate the distance to the end of the runway. If the "Δ-distance offset" parameter is not set to properly indicate the end of the runway relative to the FPAP point, the service provider encodes this parameter as "no data".

When making the approach glide path, a local vertical is introduced, which is defined as the normal to the WGS-84 ellipsoid at the LTP/FTP point and may differ significantly from the local gravity vector. The local horizontal plane is defined as a plane perpendicular to the local vertical passing through the LTP/FTP point (i.e., tangent to the ellipsoid at the LTP/FTP point). The datum crossing point (DCP) is the point at the height of the TCH above the LTP/FTP point. The FAS path is defined as the line passing through the DCP point and located at an angle equal to the glide path angle (GPA) relative to the local horizontal plane. The GPIP point is the point at which the approach path crosses the local horizontal plane. The GPIP point can really be located above or below the runway surface, depending on its curvature.

For compatibility with existing types of landing systems (ILS, MLS), the onboard GNSS/LAAS equipment should provide guidance information in the form of deviations with respect to the desired flight path defined by the FAS path. Such deviations in satellite landing systems are called "ILS-like signals".

Figure 8.5 shows the relationship between the FPAP point and the origin of the lateral angular deviation. The course width parameter and the FPAP point are used to determine the origin and sensitivity of lateral deviations. By adjusting the FPAP point position and the course width value, the desired values of the course width and the sensitivity of the "ILS-like signals" can be set. These parameters can be set so that they coincide with the course width and the sensitivity of the signals of the ILS system installed at the airfield. This may be needed, for example, for compatibility of the satellite landing system with existing means of visual landing.

The reference lateral plane is the plane that includes the LTP/FTP points, FPAP, and WGS-84 ellipsoid normal vector at the LTP/FTP point. Linear lateral deviation is the distance of the calculated position of the aircraft from the reference plane of lateral deviation. The angular lateral deflection is the corresponding angular displacement relative to the GBAS azimuth reference point (GARP). The GARP point is standardly removed from the FPAP point along the runway centerline by a fixed offset value of 305 m (1000 ft).

The sensitivity of the lateral deviation is determined by the onboard GNSS/LCS equipment based on the course width contained in the FAS data block. The service provider is responsible for setting such a value of the exchange rate parameter that gives the corresponding angle of total deviation (i.e., 0.155 DDM or 150 μA standardly used in ILS landing systems), taking into account operational limitations.

Vertical deviations are calculated by the onboard equipment relative to the GERP reference point (not shown in Fig. 8.5). The GERP point may coincide with the GPIP point (Glide Path Intersection Point with the horizontal plane) or be offset laterally relative to the GPIP by a fixed value of 150 m (corresponds to the ILS glide path positioning point). Using the offset of the GERP point, when calculating deviations from the glide path, it is possible to provide the same hyperbolic effects that exist in ILS and MLS systems and are manifested mainly at heights below 60 m (200 ft). The decision on whether to shift or not to shift the GERP point is adopted in the onboard

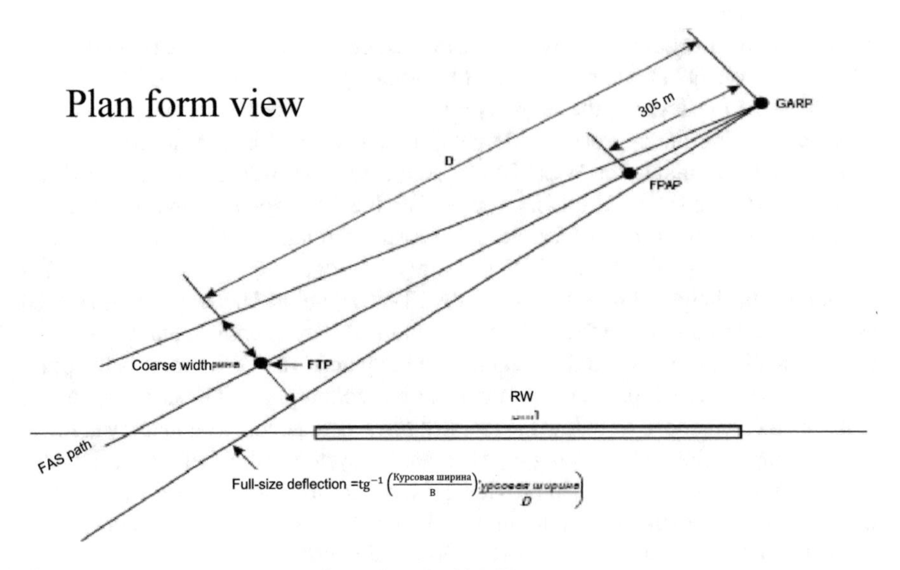

Fig. 8.6 FAS paths for approaches nonaligned with RW

GNSS/LAAS equipment in accordance with the compatibility requirements with existing onboard systems. Service providers should take into account that users can calculate vertical deviations using GERP, which is located anywhere. The sensitivity of vertical deviations is automatically set in the onboard equipment as a function of the GPA. The relation between the GPA and the vertical deviation sensitivity is equivalent to the sensitivity of the glide path offset for landing systems of the ILS type.

Some operations may require detection of a FAS path that is not aligned with the runway centerline, as shown in Fig. 8.6. For nonaligned approaches, the LTP/FTP point is not necessarily on the continuation of the runway centerline. For this type of approach, the Δ-distance offset does not make much sense and is set to "no data".

Note that the common format used for FAS data blocks is used by both augmentation systems: GBAS and SBAS. The SBAS service provider indicates in the corresponding message field which SBAS systems can be used by the aircraft that is currently using the FAS data block during the approach. The GBAS service provider may prohibit the use of FAS data along with any SBAS service. The specified message field is not used for a precise approach using GBAS signals and can be ignored by the onboard GNSS/LAAS equipment.

The service provider (LAAS) is responsible for assigning a landing identification for each approach. Landing identification does not repeat within a large geographic area. The approach IDs for a plurality of runways of a given airport are selected so as to reduce the potential for confusion and misidentification. Landing identification is published on approach charts.

Due to high requirements for accuracy of the aircraft positioning when performing ICAO Category I approaches, the FAS parameters should be detected with high accuracy: The error in specifying the LTP/FTP and FPAP point coordinates does not exceed 0.015 m. In this case, the height of the TCH is set in increments of 0.1 m, and the GPA glide path angle is in increments of 0.01°.

8.11 Accounting of Airport Location Conditions

The deployment of the LAAS includes special procedures for selecting the locations of future antennas for the VDB reference receivers and antennas. When planning the location of antennas, the requirements for minimum restrictions on the closing angles are provided.

The installation site of the LAAS reference receivers is selected in an area free from obstacles that interfere with the reception of satellite signals at the minimum possible elevation angles. In general, any concealment of GNSS satellites at elevation angles above 5° will result in a deterioration in system availability.

The design and placement of antennas for GNSS reference receivers must limit the multipath effect which interferes with the useful signal. The installation of antennas near the Earth's surface reduces the effects associated with re-reflections caused by reflections from the ground. In this case, the height of the antenna installation should be chosen to be sufficient to prevent the antenna from covering by the snow or interfering with maintenance personnel or ground transportation. The antenna should be positioned in such a way that any metal structures, such as fans, pipes, or other antennas, are outside its near-field.

At each location of the reference receiver antenna, in addition to the multipath error, the degree of correlation should also be taken into account. The reference receiver antennas are located in the places where conditions of independence are ensured in the presence of multipath propagation of radio waves.

The antenna mast should not bend under the influence of wind or under the weight of ice. The reference receiver antennas should be located in places with controlled access. Ground traffic can introduce an additional multipath error or interfere with the visibility of satellites.

The LAAS VDB transmitter antenna is placed in such a way that there is a line of sight from the antenna to any point within the coverage area for each site maintained at the FAS airfield.

It should be ensured that the VDB transmitter and the GNSS receiver antennas are spaced apart so that the maximum field strength that ensures the normal operation of the GNSS receivers is not exceeded. Providing the required coverage for several FASs in a certain airport, as well as flexibility in the placement of the VBD antenna, may require a much larger real area coverage of the transmit antenna than is required for a single FAS. The ability to provide such coverage depends on the VDB antenna location relative to the runway and the VDB antenna height. Generally speaking, increasing the antenna height may be needed to provide an adequate level of signal power for users at low heights, but it can also lead to unacceptable failures due to multipath in the desired coverage area. Optimum antenna height should be selected on the basis of analysis and taking into account the guaranteed satisfaction of the requirements for the signal power level throughout the coverage area. It should also be taken into account the influence of the Earth's surface properties, buildings, and structures on the effect of multipath.

For some LAAS arrangements, restrictions on the location of the antenna, as well as local terrain or interfering objects, can lead to terrestrial multipath and/or signal blocking. This makes it difficult to provide the required values of the field strength at all points of the coverage zone. Some LAASs may use one or more additional antenna systems arranged in such a way to create several signal paths that will collectively meet the requirements for the coverage area.

In all cases where multiple antenna systems are used, antenna operating order and message scheduling should be defined to ensure data transfer to all points in the coverage area, taking into account the specified maximum and minimum data rates and field strengths, without exceeding the capability of the VDB receiver, to adapt to power variations signal from transmission to transmission in a certain time interval. To avoid problems with data processing at the receiver associated with loss or duplication of messages, in all type 1 message transmissions for a certain type of measurement within the same frame, it is necessary to provide identical data contents.

8.12 Lateral and Vertical Thresholds of Alarm Actuation

The lateral and vertical alarm actuation thresholds for ICAO Category I precision approach are calculated as follows [1]:

The horizontal distance from the aircraft to the LTP/FTP point calculated considering the final segment of the landing path (m)	Side alarm actuation threshold (m)
$291 < D \leq 873$ $873 < D \leq 7500$ $D > 7500$	FASLAL 0.0044D (м) + FASLAL − 3.85 FASLAL + 25.15
Height of the aircraft above the **LTP/FTP point calculated considering the final segment of the landing path (m)**	Vertical alarm actuation threshold (m)
$100 < H \leq 200$ $200 < H \leq 1340$ $H > 1340$	FASVAL 0.02925H (фут) + FASVAL − 5.85 FASVAL +33.35

The parameters D and H used in these calculations are shown in Fig. 8.7.

For a precision approach by Category I, the FASLAL value does not exceed 40 m, and the FASVAL value does not exceed 10 m.

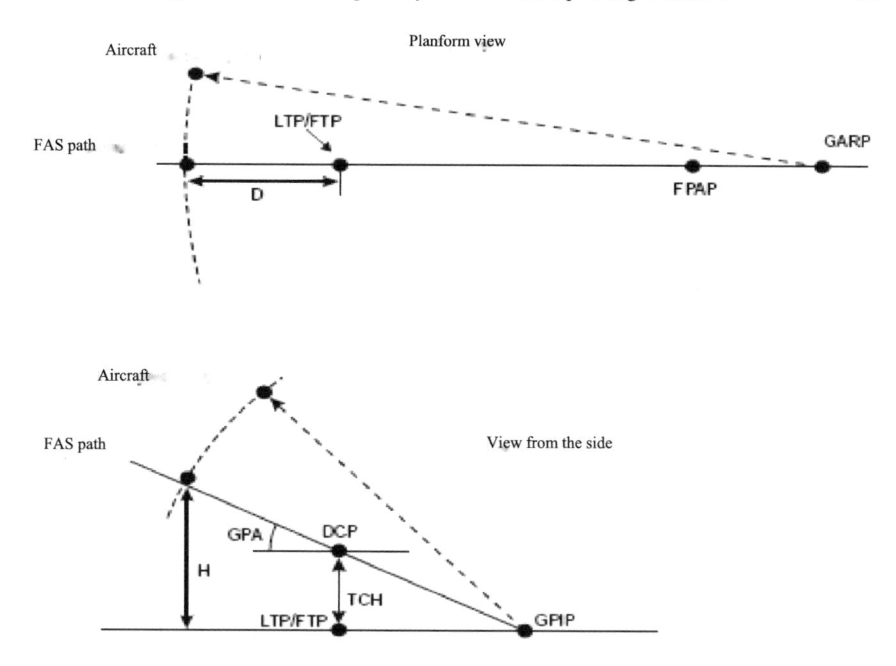

Fig. 8.7 Determination of alarm actuation thresholds

8.13 Monitoring and Actions to Keep the System in Good Operating Condition

Since the VDB signal is fundamentally important for the LAAS operation, any VDB failure, when transmitting a useful signal within the assigned time interval over the entire coverage area, should be corrected as soon as possible. Therefore, the following requirements are used as a guide for monitoring the VDB transmitter, which must be performed within max. 3 s from the occurrence time of the relevant conditions:

(a) Significant power drop.
(b) Loss of message type. Failure to transmit a scheduled message of any type. It can be a failure in several consecutive messages of the same type of transmission, or a combination of different types of messages transmission.
(c) Loss of all types of messages. Failed to send any type of message.

If there is a failure and if there is no backup transmitter, the VDB service should be terminated.

8.14 Accuracy of Setting Reference Parameters

In order to meet stringent requirements to the errors in determining the navigation parameters of aircraft, especially when performing ICAO Category I approaches, the following requirements are imposed on the relative accuracy of the geodetic survey of the LAAS reference point [1]:

- The error of the geodetic survey of the LAAS reference point relative to the WGS-84 coordinate system should be max. than 0.25 m vertically and 1 m horizontally;
- For each LAAS reference receiver, the error of the phase center of the reference antenna, should be max. than 0.08 m relative to the LAAS reference point.

Moreover, it is recommended to provide even smaller errors in determining these LAAS parameters to improve the performance of GBAS in general.

8.15 Introduction of GBAS Systems in Russia

The introduction of GBAS systems in Russia to provide a categorized GNSS landing started in 2007 and was due to the timing of the development and certification of GNSS/LAAS onboard equipment (APDD, BMS Indicator, etc.) and the start of equipping the civil airports with GBAS land subsystems. Currently, in Russia, there is only one certified ground-based LAAS subsystem—LKKS-A-2000. The appearance of the local monitoring-correcting station LKKS-A-2000 is shown in Fig. 8.8.

The photo pictures of the GNSS/LAAS equipment of the APDD and BMS-Indicator are shown in Figs. 8.9 and 8.10.

Equipment of the civil airports with ground-based subsystems of the LKKS-A-2000 type is planned to be carried out in the period by 2020 in Russian regions (at the beginning of 2018, more than 110 airports of Russia are equipped with LKKS-A-2000 equipment. It is assumed that by 2020, 100% of the fleet will be equipped with onboard GNSS/LAAS equipment, in which the positioning and landing function for GBAS according to the ICAO Category I (at the beginning of 2018, about 100 aircraft and helicopters are equipped with the onboard GNSS/LAAS equipment of type BMS Indicator).

To date, the domestic industry has developed and produced several types of onboard GNSS/LAAS equipment. All this equipment requires a separate VDB receiver (type APDD) to receive differential messages from the LAAS. Several classes of equipment have been developed. First, these are BMS-indicator and SN-4312 products, which, when interacting with APDD, perform the functions of onboard GNSS/LAAS equipment in full. Secondly, this product is BPSN-2, which does not have its own controls and must receive commands from the aircraft control system.

As an example of the interaction of advanced onboard GNSS/LAAS equipment with the onboard navigation and landing complex of the Yak-42 aircraft for imple-

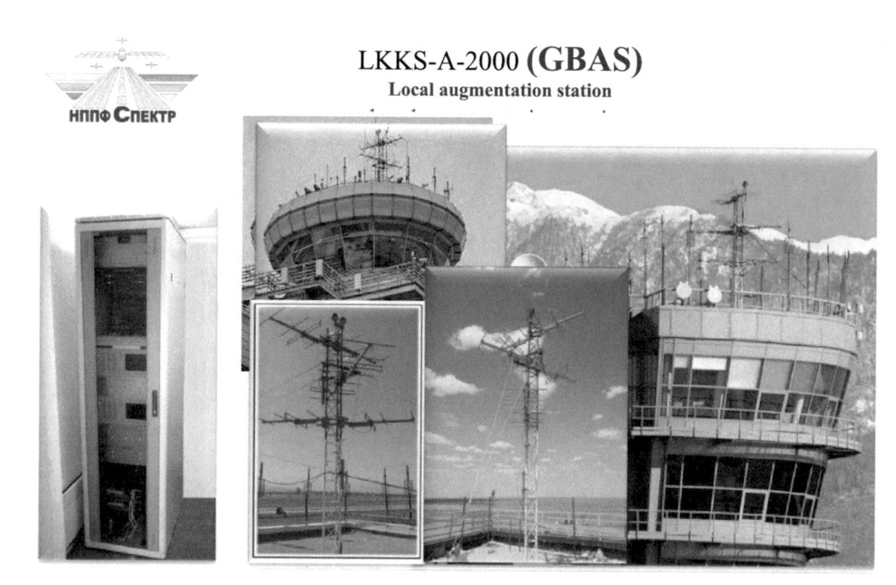

Fig. 8.8 External view of equipment and antenna-feeder system LKKS-A-2000

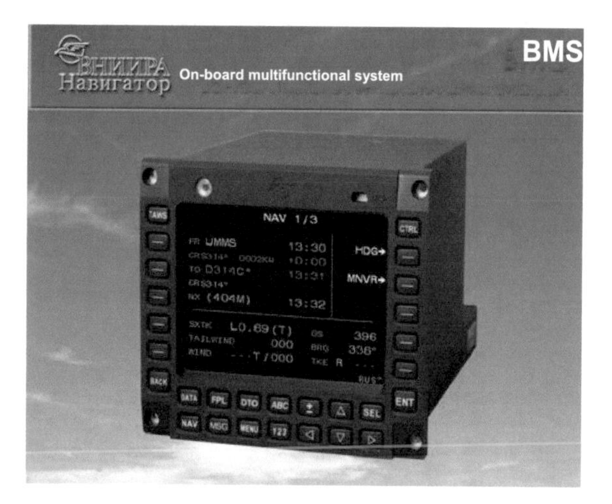

Fig. 8.9 External view of GNSS/LAAS equipment BMS Indicator

mentation of the GBAS function in Fig. 8.11. The scheme using the product BMS-Indicator is shown. It should be noted that when choosing the type of onboard GNSS/LAAS equipment for installation on the aircraft, it is recommended [7, 9] to give preference to the equipment providing reception and use of signals of satellite augmentation systems to the SBAS GNSS. In the Russian Federation, there is no SBAS system for aviation.

It should be specially noted that the implementation of instrumental approaches for landing using GBAS will improve safety and regularity of flights, reduce the mete-

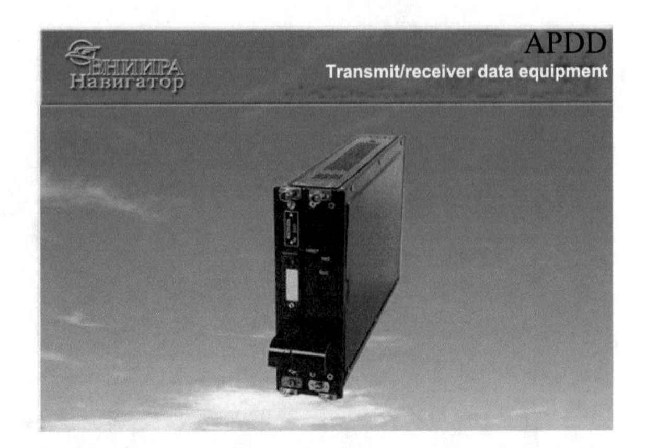

Fig. 8.10 External view of onboard GNSS/LAAS aircraft equipment APDD

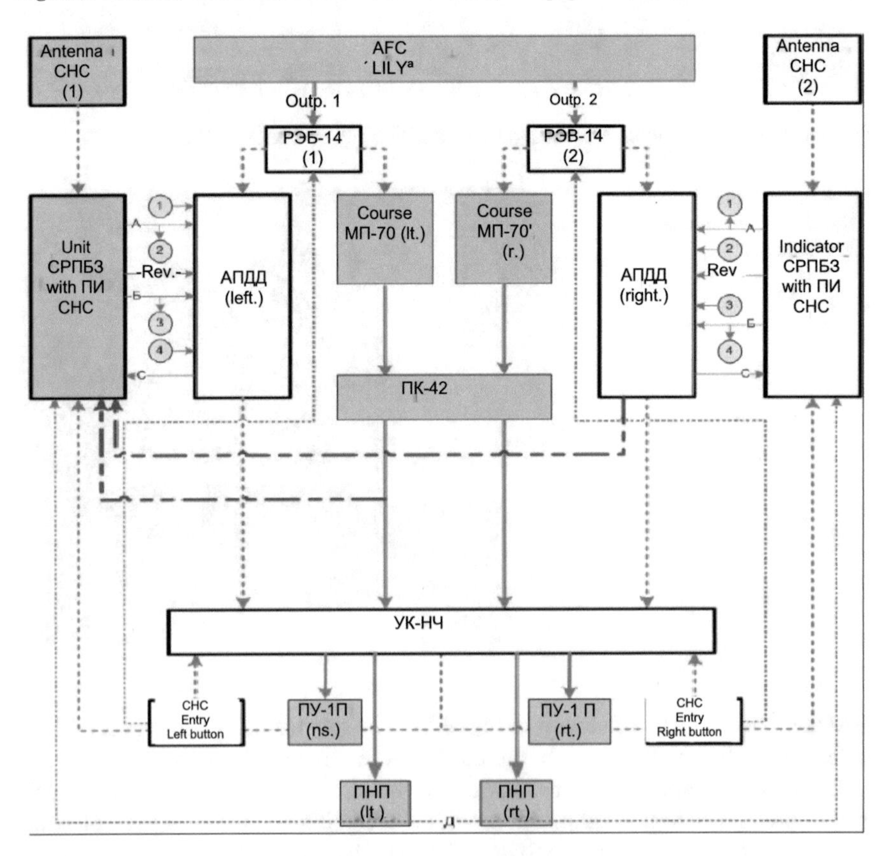

Fig. 8.11 Scheme of the interaction of the onboard GNSS/LAAS equipment with the Yak-40 aircraft equipment

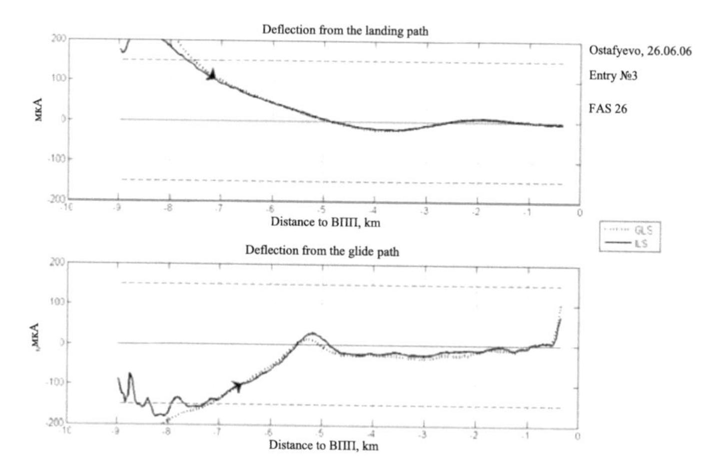

Fig. 8.12 Guidance signals from ILS and GNSS/LAAS

orological minimum of the aircraft, and it will be possible in the future to remove the traditional ground landing systems. It is extremely important that ground and aircraft GBAS equipment is much cheaper than traditional ground navigation equipment and instrument landing systems. This will significantly facilitate and accelerate implementation in the civil aviation of Russia.

Perspective onboard GBAS equipment provides the possibility of implementing the modes of director and automatic navigation. This is confirmed by the results of a comparison of the guidance signals generated by conventional systems, such as ILS, and onboard GNSS/LAAS equipment. Figure 8.12 shows the deviation signals from the landing path obtained simultaneously from the ILS system and the onboard GNSS/LAAS equipment, registered during the approach operation.

An analysis of the nature of the parameters shown in Fig. 8.12 shows that the use of ILS or GBAS-based systems for an instrument approach is practically equivalent to the pilot of the aircraft both during manual piloting and when performing calls in automatic mode using an automated control system (ACS).

The use of the GBAS system to ensure precise approach for the aircraft will be increasingly used due to its high technical and economic performance.

For aircraft pilots, the approach methodology using traditional landing systems such as ILS and MLS is practically the same as GBAS, which simplifies the practical implementation of satellite landing systems.

The implementation of GBAS is especially important for Russia, in view of the large number of unequipped airfields and the need to reduce the meteorological minimums of aircraft during flights in difficult meteorological conditions typical for many regions.

References

1. ICAO annex 10 volume 1 aeronautical telecommunications—radio navigation aids
2. Yarlykov MS (1995) Statistical theory of radio navigation. M. Radio and communication, 344 p
3. Shatrakov YG (1990) Radio navigation systems of mobile objects. M. MRP, 92 p
4. Shebshaevich VS et al (1989) Differential mode of network satellite RNS. Foreign radio electronics, No. 1, C 5–32
5. Shebshaevich VS (1992) Network satellite RNS. M. Radio and communication, 280 p
6. Olianyuk PV et al (2008) Satellite navigation systems. Spb. AGA, 98 p
7. Sauta OI et al, under editorship of Shatrakov YG (2016) Development of navigation technologies for improving flight safety. SPb, GUAP, 299 p
8. Baburov VI, Ponomarenko BV (2005) Principles of integrated onboard avionics. SPb, "RDK-Print", 448 p
9. Zavalishin OI, Korchagin VA, Lukoyanov VA, Mironov MA (2003) Monitoring of signal quality at control-correcting stations, providing differential operation of consumers of satellite radio navigation systems. Radiotekhnika 1:35–44

Chapter 9
Automatic-Dependent Surveillance

Having accurate and complete navigation information onboard the aircraft, it can be used, except for proper navigation, and to provide the surveillance function—automatic-dependent surveillance (ADS), i.e., transmit this information via an air traffic control (ATC) system for processing and displaying to the dispatcher carrying out direct ATC, and also to other consumers [1, 2].

The main advantages of ADS are the improvement of the quality of surveillance and the extension of this function to airspace areas that are outside the radar surveillance zone. ADS provides the ATC with more accurate information for the safe separation of aircraft in the air and on the surface of the airfield.

At present, the accuracy of surveillance based on radar stations (RS) is a function of range and decreases as the distance between the radar and aircraft antenna increases. Unlike the radar, the accuracy and integrity of the ADS system is unchanged throughout the service area. The surveillance system that uses the ADS will provide the ATC with the ability to accurately identify and locate aircraft that are far away from ground ATC (oceanic areas) or outside the radar (lower airspace) service area.

The availability of more accurate and reliable information on the aircraft position will enable to use the optimal flight paths and increase the airspace capacity. Besides, due to the ADS, the update of information about the aircraft in the ATC system will occur more frequently than in the radar, and thanks to this, it will be possible to track the flight path of the aircraft with higher precision. When combined with the means of accurate navigation, this will lead to the introduction of reduced separation standards, the reduction of the voice exchange between the ATC and the crew, with the release of time for working with other aircraft in the allocated airspace. All this, in the long run, will lead to an increase in the number of serviced aircraft, increase the capacity of the air navigation system.

There are two types of ADS: "contractual" (ADS-C) and "broadcast" (ADS-B) (Fig. 9.1). Data Transmission Lines used for ADS also make it possible to implement a number of other additional functions based on information exchange.

In the case of a contracted ADS, the messages (coordinates and other necessary information) are transmitted from the aircraft through the air–ground data transmission lines to a particular air traffic control unit with which the contract is established,

Sauta O.I. et al., *Principles of Radio Navigation for Ground and Ship-Based Aircrafts*,
Springer Aerospace Technology, https://doi.org/10.1007/978-981-13-8293-2_9

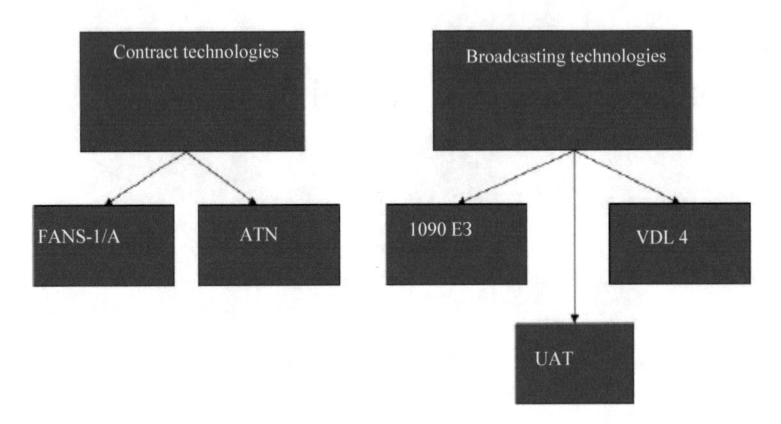

Fig. 9.1 Automatic-dependent surveillance technologies

that is, the information to be transmitted and the conditions for its transmission (transmission frequency and events initiating transmission).

At the present time, ADS-C is used mainly for servicing flights of aircraft in oceanic and remote continental regions with undeveloped ground infrastructure using satellite, shortwave and ultrashort wave communication lines and telecommunication networks.

There are two main technologies of ADS-C: FANS-1/A (ACARS) and ATN SARPS (VDL 2, AMSS).

FANS-1/A (ACARS) technology is based on satellite and ultrashort wave data transmission lines; industrial standards and specifications are published for it. The technology is recommended by ICAO for service in oceanic and remote continental areas. The equipment is installed on more than 2,000 aircraft. It is used in the ocean centers of the USA, Canada, Ireland, Australia, New Zealand, and Japan. In Russia, it is operated in the Magadan district center of the ES ATM and is planned to be implemented in the Murmansk district center of the ES ATM.

The ATN SARPS (VDL 2, AMSS) technology is also based on satellite and ultrashort wave data lines. An international standard has been published for it, but only a few dozens of aircraft are equipped with this equipment.

VDL-2 technology is supposed to serve aircraft in the airspace of Europe. Domestic ground and airborne equipment for ADS-C technologies are under development and certification.

With the broadcast ADS (Fig. 9.2), messages, coordinates and other necessary information from the aircraft are transmitted automatically in the broadcast mode to all equipped consumers (air traffic control centers, neighboring aircraft, airfield vehicles). ADS-B is used for maintenance within the range of radio visibility.

ICAO standardized three technologies for the implementation of the broadcast ADS: Extended Squitter—1090ES, VDL mode 4 and universal access transceiver—UAT.

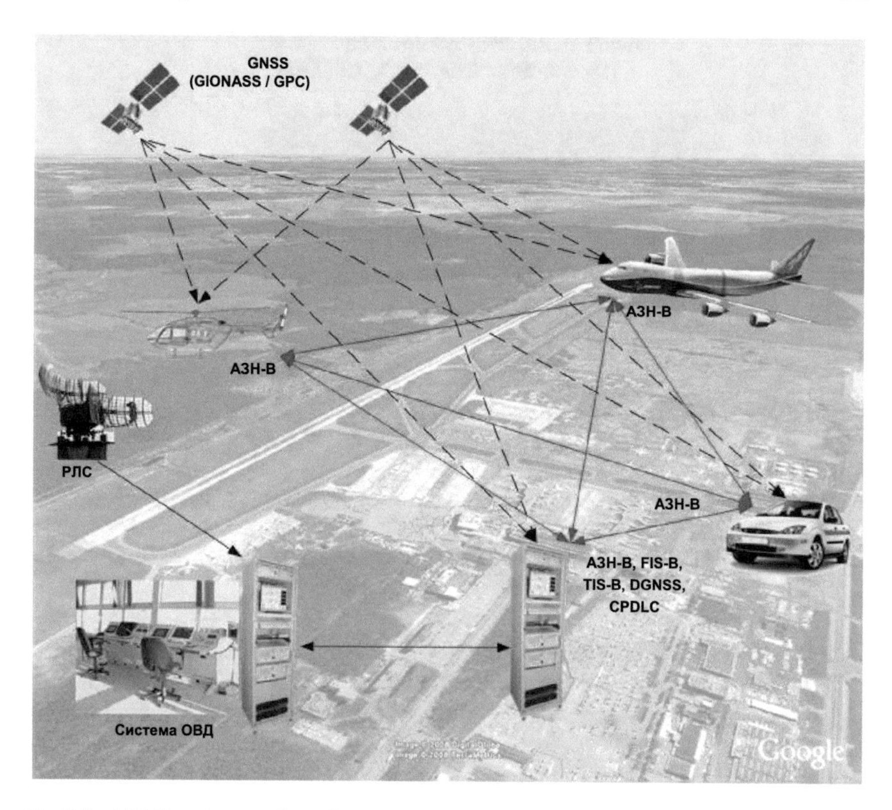

Fig. 9.2 ADS-B system configuration

As a global solution, ICAO recommended at the initial stage to use the 1090ES technology, which is the development of the SSR of the S mode. The aircraft operating in central Europe and the USA and equipped with 1090ES (Fig. 9.3) reaches 80–90%. Taking into account the level of equipment of the aircraft, it is assumed that in Russia it will be used for flights in the upper airspace and on international air routes. This technology allows the transfer of data from the aircraft through ground-based receiving stations to air traffic controllers to monitor the movement of aircraft (OUT mode). The introduction of ground receiving stations will significantly expand the surveillance area of the ATC system and create a reserve in the event of secondary radars failure.

The implementation of the 1090ES equipment onboard the aircraft is carried out by upgrading the TCAS collision avoidance systems, which are mandatory for flights in the international airspace. The results of observations in the Moscow air zone (Fig. 9.4) showed that 40–50% of the aircraft are equipped with 1090ES equipment. Ground-based equipment is the development of secondary S-mode radar systems, but does not require the emission of powerful interrogation signals and is therefore much less energy-intensive and less expensive to operate. According to Eurocontrol,

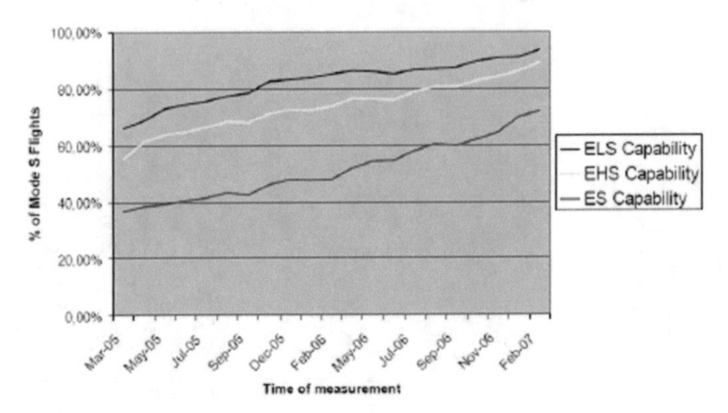

Fig. 9.3 Equipping aircraft with Mode S technology (according to the Charles de Gaulle airport in France, Paris). ELS—basic surveillance based on SSR; EHS—improved surveillance using SSR Mode S; ES—Extended Squitter for ADS-B implementation

operational use (75% coverage of ground and airborne infrastructures) of 1090ES OUT for surveillance in Europe will be achieved in 2015 (2014—USA) with the first applications in 2008–2009. The organization of data transfer in the reverse direction via the air-to-air link (IN mode) is assumed in the future. International standards for these functions are under development. The use of 1090ES IN for air-to-air and ground-to-air communication is planned to start after 2010 (in the US—after 2020).

It should be noted that the conditions of flights and the aircraft themselves operating in the lower airspace are significantly different from the long-range aircraft. Therefore, the US decided to use for the ADS-B in the lower airspace a separate UAT technology with a single receive-transmission frequency of 981 MHz. That is, in the US it is planned to use two technologies: In the upper airspace—1090ES; in the lower airspace—UAT. The strategy for introducing ADS-B in the US has been published and is being refined in consultation with airspace users. The results of the discussion revealed a number of contradictions between airspace users, industry and air navigation service providers. In particular, it is necessary to solve the question of a very high cost of the dual-mode solution and the advantages that airspace users will receive from its implementation. Currently, regional projects use UAT technology (Alaska). VDL4 technology is not supported by the world aviation community (including the Swedish aviation administration).

The technology is based on the application of the VHF band data transmission lines, which allows for two-way data exchange between ground and air–ground. In this regard, the used communication line allows information exchange between the aircraft, between the aircraft and the airline, to organize a digital transmission of data "ATC-pilot", to transmit information about the air situation and meteorological information from the ground to the aircraft. However, ICAO requirements for the integrity of the VHF data link are not developed and the systems can not be certified.

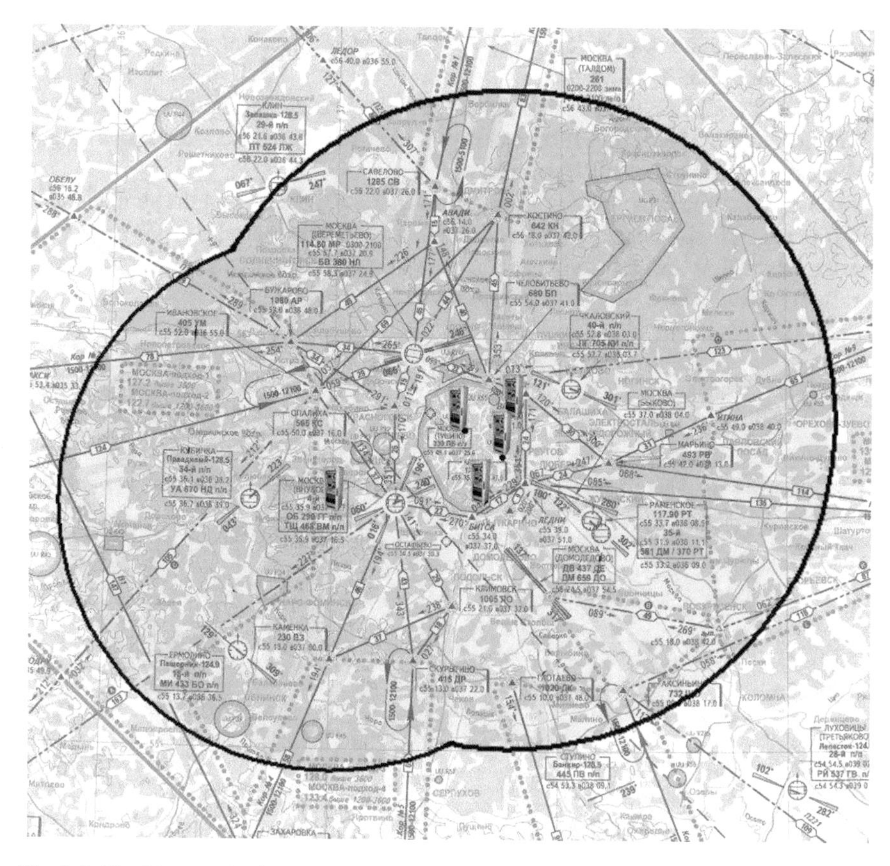

Fig. 9.4 The Moscow-ADS project in 2007

Thus, the global compatibility of the Russian air navigation system will be implemented through the use of 1090ES technology (along with the use of other specialized airborne equipment).

The FAA recognizes that there are vulnerabilities in the use of the GPS system as a means of locating the aircraft. There are cases when GPS may not be reliable in certain places and at certain times due to the presence of interference from other radio equipment or solar activity. In this case, a backup strategy is required that provides ATC with surveillance information. The strategy should provide the monitoring function for the ATC to the same extent that modern backup types of surveillance provide. In other words, at least the same level of capacity should be retained with the loss of the satellite signal, as would be the case with the loss of the radar signal.

This approach is most consistent with the strategy of preserving the reduced radar network. According to this strategy, radar services will be provided in the approach zone with high and, in part, medium traffic density, as well as in the airspace on the route at an altitude of more than 5386 m (18,000 ft). To do this, it will be necessary to maintain a certain part of the SSR for a long-term perspective, but at the same time to reduce their total number.

The services of SSRs will be maintained at strategic route positions and throughout the airfield space where they are currently provided, in order to compensate for single-unit failures in the onboard electronics.

The introduction of new CNS/ATM technologies is a priority for the air navigation development; it will bring significant benefits to both the provider of air navigation services and, first and foremost, to airspace users. This task should be addressed jointly by the ATM system and the operators.

References

1. ICAO Annex 10 Volume IV Aeronautical telecommunications—Surveillance and collision avoidance systems
2. Yarlykov MS (1995) Statistical theory of radio navigation. M Radio Communication 344 pp

Printed in the United States
By Bookmasters